Nutrition of white-tailed deer

PRESENTATION

The white-tailed deer (Odocoileus virginianus) is a wild animal that inhabits practically the entire American Continent. Economically, it is considered the most important hunting trophy in the world. White-tailed deer habitat is found in virtually all habitats, provided these provide sufficient shelter and food. It is not common in the drier and more open parts of the xerophilous scrub or in the more dense and humid parts of the tropical evergreen forest. It is distributed from the center of Canada to Bolivia. In Mexico, practically in all the country, except in Baja California state. The feeding potential of browsing native shrub plants and native grasses and grasses, which form the main source of food for white-tailed deer, is practically unknown by many holders of land. In some regions, shrub control practices were carried out until almost complete eradication, perhaps because the first investigations that were carried out on the subject, considered only the improvement of the grassland with the sole objective of providing greater forage production for domestic livestock, forgetting about the importance of wildlife and their diet. Generally, pasture management practices have been focused on completely dismantling the area to implement grass pastures such as the buffel (*Cenchrus ciliaris* L.), which has brought serious consequences, since on the one hand the soils have been eroded and another, the habitat of wildlife has been destroyed.

It is not easy to obtain precise estimates of the nutrient content in the diets selected by the white-tailed deer under natural conditions, due to the impossibility of observing the deer's preferences for the different plants that occur in the pasture. However, the deer, like any living organism, requires a series of nutrients in sufficient quantities to develop its vital functions. Few studies have been reported to express their nutritional requirements, since in their natural state it is difficult to detect their needs. The scientific reports of deer nutritional requirements have been carried out with deer in captivity, although in a certain way the results of these studies can be inferred from wildlife conditions. Of the basic characteristics necessary for good management of white-tailed deer, nutrition represents the most important factor. Without proper nutrition, the deer could not develop its full genetic potential. Practically 50% of deer deaths are associated with poor nutrition. Malnutrition is associated with small bodies, poor development of antlers, poor reproduction and poor survival of fawns. This text is presented as a need to provide scientific literature, on Nutrition of Whitetail Deer, to students, scientists and scholars of nutrition and feeding of wildlife, with accents on ruminal physiology, nutritional ecology, habitat, water, protein compounds, carbohydrates, lipids, energy, secondary compounds of plants, minerals and vitamins.

Profesor: Roque Gonzalo Ramírez Lozano, Ph.D.
Universidad Autónoma de Nuevo León,
Facultad de Ciencias Biológicas, Dpto. Alimentos,
Ave. Pedro de alba esquina con Manuel Barragán, S/N,
San Nicolás de los Garza, Nuevo León, ZC. 66455, México,
Ce: roque.ramirezlz@uanl.edu.mx

CONTENT

	Page
Chapter 1. Biology and Behavior	1
Chapter 2. Digestive Physiology	15
Chapter 3. Water	34
Chapter 4. Habital of white-tailed deer	41
Chapter 5. Proteins	51
Chapter 6. Carbohydrates	72
Chapter 7. Lipids	86
Chapter 8. Energy	99
Chapter 9. Minerals	107
Chapter 10. Vitamins	123
Chapter 11. Lignin	137
Chapter 12. Tannins	142
Chapter 13. Voluntary intake	156
Chapter 14. References	170

This book is dedicated to:

To my wife: Emma and my daughters: Nancy, Sylvia y Brenda.

Chapter 1

Biology and behavior

> **Abstract**. The white-tailed deer represent the largest group of mammals in North and Central America. It is known mainly 38 subspecies of white-tailed deer: 30 subspecies for the northern and central part of the continent and 8 subspecies for the southern part. The South American subspecies are clearly distinguished from the North American ones by genetic differences, smaller antlers, absence of metatarsal gland and lower weight and body size. However, due to repopulation programs, which has been carried out in recent decades, it is suspected that there have been cross-breeding between the different subspecies. The characteristics of the reproduction of the deer are characteristic of the species and depend on the conditions of the habitat. Also during the mating season, the females secrete hormones and pheromones that tell the deer that the female is warm and ready to mate. The horns or antlers of the deer are bones or bone (temporal) extensions that develop from the frontal bone of the head of the animal, but with the particularity that they are not covered or sheathed as in the case of the horns. White-tailed deer age or life expectancy is influenced by hunting pressure and other mortality factors. The coloring of the white tail deer helps in its camouflage, thermoregulation and in its communication. The deer is twilight in its pattern of behavior. It is a species that moves through trail systems that lead to stalls, feeding areas and escape routes, where it is common to observe traces and feces. The behavior of males of a wild population of white-tailed deer during the "run" is directly associated with the age of the animal. Sexual segregation is a recurrent behavioral pattern in which it is observed that the adult male and female members of a population are aggregated in the mating season and segregate outside the time of heat (gestation and births) or, forming groups either with their same sex, with a greater predominance of one of the two sexes and with juvenile individuals. The concentration in the serum of another thyroid hormone, thyroxine (T4), has been associated with the consumption of protein in the diet and with the concentration of energy.

Introduction

White-tailed deer represent the largest group of mammals in North and Central America. Accurate estimates of its number have not been made, but probably, in the entire American continent, represent between 8 and 15 million individuals. Even though, in many areas, their number had decreased, until almost their extinction, they have recently reached a significant number due to the improvement of their habitats and the development of a hunting culture. Whitetail deer, as we know it, probably takes more than 500,000 years without changing its phenotype, its ancestors have populated the American continent for almost 20 million years, the constant interaction between it and its natural environment was slowly influencing and modifying to both. In such a way that, over time, the siege of its predators that stalked and killed it, shaped its slender body and improved its senses, transforming it into a fast, strong and careful animal. As he always evolved like a prey, he sought protection in the most rugged and distant lands, his movements became cautious, his acute senses and perception of what surrounds him were exacerbated, making him a stealthy and distrustful individual. Also, his digestive physiology was adapted to consume forages that were within reach, but that mostly contain toxic substances or secondary compounds.

Cervids of the world

Cervids (Cervidae) are a family of ruminant mammals that includes deer or crvids. They have thin legs, split hooves and long necks. They are slender herbivores. They have straight or speckled hair, hooves with two fingers, and are the only mammals that grow horns or new horns every year, formed by dead bone. Only adult males develop them and use them during mating season, when cervids compete for females. The antlers begin to form from two protuberances of the skull. When growing, a velvet covers them. When the antlers grow, they begin to branch out. Finally, the velvet falls. Thus the antlers is complete. They inhabit several areas of the planet, so they can be found in Europe, Asia, America, North Africa and some Arctic areas. It was introduced by man in New Zealand and Australia.

The deer vary in size, with the elk being the largest, and the venadito or pudú of South America, the smallest. Most deer have a gland near the eye that contains pheromone, a substance that helps them to mark their territory. Males use this substance when they are upset by the presence of other males. Most deer species live in family groups around a female, although there are others, such as the musk deer, which lives as a couple. They feed on leaves, branches and shoots of plants. The period of gestation of the females varies between 160 days to 10 months according to the species; They give birth to one or two babies a year.

Taxonomic classification and distribution

The white-tailed deer taxonomically is classified as follows:

Order: Artiodactila (ungulates)

Suborder: Rumiantia (ruminants)
Infraorder: Pecora
Superfamily: Cervidae (cervids)
Genus: *Odocoileus*
Species: *virginianus*

Artiodactyls or ungulates are characterized by having a foot with a helmet or hoof with even fingers. In addition, the fact of belonging to ruminants, places it as a mammal that ruminates and that has a stomach divided into four compartments (rumen, reticulum, omasum and abomasum) and that has no upper incisors. In addition, it is a very selective browsing herbivore. Because it has a high birth rate, a wide distribution of young animals and can tolerate high and low temperatures very well, the white-tailed deer is very widely distributed throughout the Americas, from where the *Odocoileus* genus originates; of which 38 subspecies of white-tailed deer are known: 30 subspecies for the northern and central part of the continent and 8 subspecies for the southern part. In Mexico it is estimated that there are 14 of the 30 subspecies of those reported in the north and center of the continent, that is, 47% of the subspecies that exist from Canada to Panama. However, the white-tailed deer is the only type of deer that has had the ability to be distributed over most of the Mexican territory, except for the state of Baja California, Mexico.

Below are most of the subspecies of white-tailed deer, the authority that identified them and their place of location within the American continent.

1. *Odocoileus virginianus* borealis (Miller 1900) Eastern Canada and northeastern United States.
2. *Odocoileus virginianus* dacotensis (Goldman & Kellog 1940) North and South Dakota, Nebraska, Wyoming and Southwest Canada.
3. *Odocoileus virginianus* virginianus (Zimmermann, 1780) Virginia.
4. *Odocoileus virginianus* macrourus (Rafinesque 1817) Kansas.
5. *Odocoileus virginianus* mcilhennyi (Miller 1928) Avery Island, Louisiana.
6. *Odocoileus virginianus* taurinsulae (Goldman & Kellog 1940) Bull Island South Carolina.
7. *Odocoileus virginianus* osceola (Banqs 1896) Florida Coasts.
8. *Odocoileus virginianus* seminolus (Goldman & Kellog 1940) Interior of Florida.
9. *Odocoileus virginianus* clavium (Barbour & G. M. Allen, 1922) Florida Keys
10. *Odocoileus virginianus* ochrourus (V. Bailey 1932) Rocky Mountains

11. *Odocoileus virginianus* leucurus (Douglas, 1829) Columbia River, Oregon and Washington states
12. *Odocoileus virginianus* couesi (Coues & Yarrow 1875) Arizona, southeastern California, New Mexico, and northwestern Mexico
13. *Odocoileus virginianus* texanus (Mearns 1898) Texas, Oklahoma, southeastern Colorado, northeastern Mexico.
14. *Odocoileus virginianus* carminis (Goldman & Kellog 1940) Northern Mexico.
15. *Odocoileus virginianus* miquihuanensis (Goldman & Kellog 1940) Central Mexico.
16. *Odocoileus virginianus* mexicanus (Gmelin 1788) Puebla, Mexico.
17. *Odocoileus virginianus* acapulcensis (Caton 1877) South of Mexico.
18. *Odocoileus virginianus* veraecrucis (Goldman & Kellog 1940) western of Mexico.
19. *Odocoileus virginianus* thomasi (Merriam 1898) Oaxaca and Chiapas.
20. *Odocoileus virginianus* yucatanensis (Hays 1872) Yucatan.
21. *Odocoileus virginianus* nelson (Merriam 1898) Guatemala.
22. *Odocoileus virginianus* truei (Merriam 1898) Centroamérica
23. *Odocoileus virginianus* chiriquensis (J.A. Allen 1910) Panama.
24. *Odocoileus virginianus* rothschildi (Thomas 1902) Coiba, Panama.
25. *Odocoileus virginianus* curassavicus (Hummelink, 1940) north of Colombia, Curazao.
26. *Odocoileus virginianus* goudotii (Gay & Gervais 1849) Andine of Colombia and Venezuela.
27. *Odocoileus virginianus* margaritae (Osgood 1910) Isla Margarita, Venezuela.
28. *Odocoileus virginianus* apurensis (Brokx, 1972) Llanos colombo-venezolanos and northestern of Amazonia.
29. *Odocoileus virginianus* ustus (Trouessart, 1913) Andine Zone of Ecuador and south of Colombia.
30. *Odocoileus virginianus* tropicalis (Cabrera 1918) The Pacific region of Colombia.
31. *Odocoileus virginianus peruvianus* (Gray 1874) Andes of Perú
32. *Odocoileus virginianus* gymnotis (Wergmann 1833) Venezuela, Guyana and Surinam.
33. *Odocoileus virginianus* cariacou (Boddaert 1784) Guayana Francesa and north of Brasil.

The South American subspecies are clearly distinguished from the North American ones by genetic differences, smaller antlers, absence of metatarsal gland and lower weight and body size. For these reasons some experts have proposed to classify them into two different species and give the South American species the name of *Odocoileus cariacou* Boddaert

According to the current international record books of the most important trophies ("Boone and Crockett Club" and "Safari Club International"), only three of the 14 Mexican white-tailed deer subspecies are susceptible to classify and enter them; and it is for this reason that they have the best population densities and receive greater protection from farmers and ranch owners. These subspecies are:

1. *Odocoileus virginianus* texanus, popularly known in Mexico as "Texan", is located mainly in the northeast of Coahuila, north of Nuevo León and northwest of Tamaulipas,
2. *Odocoileus virginianus* couesi popularly known in Mexico as "coues", is located mainly in the states of Chihuahua and Sonora
3. *Odocoileus virginianus* carminis. popularly known in Mexico as "deer of the carmen", is located in the mountainous areas of northern Coahuila (only recognized in the record book of Safari Club International).

However, due to the repopulation program that has been carried out in the last decades, it is suspected that there have been crossbreeding between the different subspecies of deer that have been distributed historically in that region. when carrying out a study to determine the genetic variability among the four subspecies of white-tailed deer of Northeast Mexico (O. v. carminis, miquihuanensis, texanus and veraecrucis) using the technique Sequence Scanning Base Excision (BESS-T), a part of the mitochondrial control region (D-loop) in 106 DNA samples obtained from individuals of these subspecies and based on the BESS-T patterns, haplotypes were defined and the interpopulation genetic variability among subspecies was evaluated. To this end, 106 samples of DNA obtained from blood and hair were analyzed and interpopulation genetic variability among subspecies was evaluated.

The genetic constitution of the animals (genotype) was evaluated through the maternal line thus obtaining half of the genotype (haplotype) and 24 haplotypes were found: of which 17 were presented in *O. v.* texanus, of which four of them are shared with the other subspecies, this being an indicator of crossbreeding among the subspecies. Three haplotypes were only presented in *O. v.* veraecrucis, two in *O. v.* carminis and two in *O. v.* miquihuanensis which indicates that these populations in the study regions have remained without cross-linking among themselves or with the subspecies Texanus. In the subspecies *O. v.* texanus inter and intra population variability was evaluated by subdividing it into three populations according to their geographical origin (Coahuila, Nuevo León and Tamaulipas). Only two haplotypes (1 and 5) are shared in the three entities, which coincides with the repopulation programs that have been carried out in the three states favoring the mixing of the genetic material within the *O. v. texanus* It is concluded that the genetic diversity index for Coahuila, Nuevo León and Tamaulipas is 0.82, 0.56 and 0.89 respectively, which indicates that populations of white-tailed deer have introduced genetic material in places where they did not exist historically.

Mating and reproduction

White-tailed deer do not have gregarious habits; however, it is common to form small groups of four to six females, two or more females and their offspring, or, that two or more adult or young males are grouped during the times that do not correspond to the mating season (December and January). The behavior of males in groups of different ages, is common to observe practically from the months of February and March, until the month of October and even November; being able to report cases of two or more males of different ages together during this time. During the mating season, which generally corresponds to the months of December to February, males tend to keep separate.

The characteristics of the reproduction of the deer are characteristic of the species and depend on the conditions of the habitat. Active sperms have been found in males from September to March of the following year. Likewise, in the autumn an increase in ovarian activity has been observed, in which there is development of one or more mature ovules. The females have a heat that lasts approximately 24 hours and, if they are not covered, return to heat once or twice more with intervals of 28 days, depending on the quality of the habitat, the year in question and the physical condition of the animal. It has been calculated that an adult male (4 to 5 years old) can cover, in a good habitat, an average of five or six females per year. However, in confinement; in intensive breeding sites, an adult male (which does not have competition from other males) can cover up to 15 females or more. Most of fawns are born in May and June, in these latitudes, after a gestation period of 187 to 222 days. The gestation is prolonged when the habitat is nutritionally poor. The adult females usually give birth when the conditions of the pasture are favorable and, therefore, provide adequate nutrition. However, when in some years the quality of the diet could be nutritionally poor because the spring rainfall was low, the females give birth to a single fawn.

Glands in deer

Deer possess many glands that allow them to produce essences, some of them are so powerful that they can be detected by the human nose. The five types of glands are 1) the orbitals, 2) in the front of the head, 3) the tarsals, 4) the metatarsals and 5) the inter-digitals. The orbital glands are located around the eyes and release the essence by rubbing the head normally the area around the eyes on the hanging branches. The ones on the front of the head are between the eyes and the antlers. The tarsal glands are found in the lower outer part of the hind legs. The essence is released when the deer rose vegetation when walking. The metatarsal glands are found in the inner part of the knees of the hind legs and are the most powerful. During the mating season, the deer urinates in such a way that it drains into the inside of its legs to where the metatarsal glands are located. And the interdigital glands are located between the front nostrils. Once here he rubs them, carving the urine with thick hair and mixing the secretions of the glands, urine and bacteria to produce a strong smell. Also during the mating season, the females secrete hormones and pheromones that tell the deer that the feme is ready for mating.

Marking the territory is the most obvious way in which the white-tailed deer communicate. Although the deer are the ones that mark the most, the females visit these places frequently. One way to mark is to carve, this is achieved when the deer carve with its bark the bark of trees of small diameter, thus marks its territory and polishes it horns. Also, to mark their territory, the deer scratch the ground with their hooves, usually following a pattern in line.

Antlers and its function

The antlers of the deer are bones (temporal) extensions that develop from the frontal bone of the head of the animal, but with the particularity that they are not covered or sheathed as in the case of the horns. The antlers are molted and regenerated every year, something that does not happen with the horns, and during their development and formation, they really are a relatively soft living bone; however, once their growth is over, they have a very solid consistency and are really a dead bone (which will no longer continue to grow). They are formed of approximately 43% protein and the rest of minerals (mainly Ca and P). The antlers begin their growth in May and end their growth in September. During the whole stage of its development (approximately 100 to 120 days), the new antlers remain covered by a very thin and sensitive membrane which is covered with hair (grayish brown) and whose texture is common called velvet.

The antlers are a fascinating component of the anatomy of the deer. Typically, only adult males possess them. This is since the male hormone testosterone is the main metabolite responsible for regulating the growth of antlers. It is believed that the most important function of antlers in males is to determine the success of their mating, because the hardening of the antlers occurs precisely in the mating period. The antlers, in fact, constitute a symbol of the position or hierarchy that the males keep within the population to which they belong, and they achieve it based on fights between them, and their size and hardness are directly associated with genetic factors, age and condition. nutritional status of their habitat. The antlers increase in size (at one point) as the deer ages. Once the mating period ends, the antlers deteriorate producing necrosis, weakening at the base, which causes its fall, approximately between December and February.

One of the most controversial issues in the management of wild populations of white-tailed deer is the question of whether males should be eliminated from the population with pointed horns, commonly known in the region as aleznillos and whose name obeys its similarity with the pointed awls that shoemakers use to pierce, sew and stitch the skins. In this regard, it is important to refer the studies carried out in Texas, USA, which showed that the deer of a year and a half old, whose first antlers were of the awl type or aleznillos and that were supplemented with a diet based on 16% of crude protein, achieved at the age of 2.5 years an average of 6.78 peaks per deer and an average opening of maximum width between antlers of 29.5 cm (11 5/8 ".) These same deer, achieved at the age of 3.5 years, an average of 7.22 peaks per deer and an average opening of maximum width between antlers of 36.2 cm (14 1/4 "). According to the above, it can be established that a large percentage of deer whose first antlers (1.5 years of age) are of the awl type, with adequate nutrition and high percentage of crude protein, can develop antler poles in the future. better features; however, they will hardly become good trophies, since according to the study referred to, the males whose first antlers (1.5 years old) were not of the awl type, that is, males with baskets of antlers of four or more peaks, achieved a greater average of peaks and of maximum opening between antlers. Being the most significant, the fact that their antlers were on average 71% heavier during the two years that the research lasted, which can be interpreted as antlers of greater volume.

Therefore, the growth of antlers has a genetic basis, but is influenced by the environment. To obtain the best development of the antlers, the herds must be managed to obtain the best possible genetics and simultaneously the habitat must be managed to obtain the best possible nutrition. In this way it will be ensured that deer express their greatest genetic potential. The characteristics of the basket can be improved by selecting those animals that have good basket even in adverse conditions, contrary to those animals that only develop good baskets under good habitat conditions. Therefore, the best selection of the animals should be during nutritional stress (drought), in these conditions, the animals will appear and may be removed from the herd.

Life expectancy

White-tailed deer age or life expectancy is influenced by hunting pressure and other mortality factors. In free form the deer teeth after the seventh or eighth year of life, is so worn that it is possible that it dies by starvation, or that the lack of adequate nutrition makes it easy prey for predators, diseases or parasites However, evidence has been found that there are deer with an approximate age of 18 to 20 years, although a male deer that survives more than 6 years represents a small proportion of all males in regions where hunting is intensive. However, it has been found that when the habitat conditions are good and proper management has been done, the 6.5 and 7.5-year-old deer reach a good corporal physical condition and a good development of their antlers. Quality that tends to decrease after 8.5 years of age, due to the consequent wear of their molars, which results in a low use of the forage consumed for their food.

Physical appearance

The coloring of the white tail deer helps in its camouflage, thermoregulation and in its communication. In adulthood, they change their coat twice a year 1) in the summer the hair is reddish-brown, and the coat is very thin and light to help the deer stay cool and 2) during the winter it is dark grayish, brighter and longer, in addition to being a better insulator against low temperatures. The speckling of the fawns' hair tends to disappear in a period that can vary between 70 and 100 days after birth. Generally the fawns are born in a dense thicket, where it is possible to hide them from their predators, and although the newborns may seem easily vulnerable by coyotes or other predators, the protection provided by their mottled fur, makes them go virtually unnoticed when confused with the shadows that the vegetal cover projects and, by its almost total lack of smell during the first days of births.

The darker colored antlers are common in older male deer, older than 4.5 years of age, because velvet is composed of a greater number of veins and capillaries, which results in a darker spotting effect of hemoglobin. In younger males, antlers with lighter coloration are common, because there are fewer veins and capillaries in the velvet. There is a great variability in the complexion of white-tailed deer, largely due to genetics, sex, age and nutritional status. In general, it has been reported that it has a size of medium to large, males weigh within a range of 34 to 180 kg and females of 22.5 to 112.5 kg. Its height varies from 91 to 107 cm.

Behavior

The deer is twilight in its pattern of behavior. The above means that it is more active during the dawn and during the darkening. As an adaptation to this lifestyle in scrubland where visibility is limited, the deer has very well developed the senses of smell and hearing. The sense of smell is vital for your safety, feeding, growth and mating. The sense of hearing is also extremely important to detect the danger and to communicate with other deer. The scope of his sight is important for his survival, his eyes are designed to detect movements and he has a proportion between canes and cones that increase his ability to see with low levels of light. Like many mammals, it is believed that deer do not detect colors, colors detect them as shades of gray.

It is a species that moves through trail systems that lead to stalls, feeding areas and escape routes, where it is common to observe traces and feces. The feces are constituted mainly by vegetal material of variable size and shape that normally do not exceed the 1.5 cm of length, although they can be loose or compacted in bulks of greater size. The tracks of the front and back legs are practically the same size and usually measure 5 to 6.5 cm long by 3 to 5 cm wide, normally only the two central hooves are marked, however, during the race, on steep slopes and on soft ground can be observed two small fingers called false hooves. The pitches are usually densely vegetated places where it is possible to take refuge and rest, bushes are often found browsed to a height of 1.5 m and bark eaten at the same height. Another common trace is generated when the males, at the end of the development of the antlers and lose the velvet that cover them, carve their antlers against small trees and shrubs, which is marked in the bark in a stretch of approximately 50 cm.

The behavior of males of a wild population of white-tailed deer during mating (bullfight) is directly associated with the age of the animal. The dominant males of 4.5 years of age or older, are deer that defend the female that they find in estrus within a certain area, expelling any male that pretends to approach her. In contrast to the above, younger, non-dominant males (ages 1.5, 2.5, and 3.5 years old) are common in wandering from one place to another in search of estrus females. It is common for dominant males to define lines of 100 or more meters in length within their home environment, where at each 30 or more meters, they prepare sites called diggings, which they prepare by stamping the ground with their front legs. They complement the marking of these sites with the urine that they deposit on the digger, which they trample to mix it with the ground. These diggers are checked by the males once or twice a day (if they are not accompanying a female that is going to enter estrus), in order to know if they were visited by a female in estrus or close to being (as these urinate on the diggers when they visit them), which is detected by the male by the presence of a higher concentration of sex hormones in the urine deposited. If this happens, the male goes in search of the female in estrus to cover it, following the trail that it leaves when walking.

There is a recurrent pattern in which it is observed that the male and female adult members are added during the mating season and live separately (segregate) outside the time of heat (gestation and births) or, forming groups (either with the same sex, with greater predominance of one of the two and with juvenile individuals); this different use of space has been called sexual segregation. Sexual segregation has been widely studied among mammals with marked sexual dimorphism and has been frequently associated with allometric and physiological differences, which can cause a different selection of habitat, diet and sex activity patterns. The sexual dimorphism in mammals produces as a common feature, differences in body size between the sexes, these are clearly reflected in all the species of suborders Ruminantia.

Sexual dimorphism is the result of sexual selection and is more evident in polygamous species, since there is a positive correlation between body size and mating opportunities in males. Among the groups of mammals that present a high degree of sexual dimorphism are the cervids of temperate zones, where the variations in the size of the body go from 20 to 70% of the corporal mass between sexes. In the case of white-tailed deer (*Odocoileus virginianus*), in some of their northern populations, the body mass of females can be 49.5 kg, while the male can weigh 82 kg being the percentage difference between body mass of 65.66%.

Deer have specific habitat needs, so their distribution and abundance are limited to an area given by the quantity, quality and heterogeneity of available resources. These characteristics have an important influence on the size of the household environment, which, in turn, is determined by factors such as: population density, quantity and quality of the diet, plant cover, presence of water sources and reproductive activities. So, it is expected that the environmental and reproductive conditions modify the size of the household environment. It is important to consider that the use of an individual's habitat is a complex process, which shows variations in the preferences of both sexes from one year to the next and even when both sexes are in the same physical environment, intersex differences in habitat preference are a phenomenon not yet deciphered.

Therefore, sexual segregation in ungulates such as white-tailed deer is given by social, spatial, temporal and physiological factors, such as: population density, social organization, mating system, distribution of resources, environmental conditions, risk of predation, body size, differences in diet selection and digestive function among ruminants. These factors could explain much of the variation in habitat selection.

Young deer roam within their family ranges. These movements are known as transfers. Forage conditions and coverage within the family can vary temporarily due to drought or other factors and under these conditions deer can move to find sites with better forage or cover conditions that could be used temporarily. Disturbances caused by humans or lack of water can force deer to move to areas with less disturbance and greater water availability. When the deer remain in the new area where they have moved, this is known as dispersion and in these cases the deer do not return to their original family environment. It has been showed that in a population of white-tailed deer, males less than 3 years of age are those that disperse more frequently and males of the year disperse more than any category of age or sex within the population, being those who can separate from their mothers and start exploration trips when they are approximately one year old. In the western plains of southern Texas, half of the yearlings in the deer population dispersed on average 4.25 km from their original family range.

The Jacobson organ, also known as the vomeronasal organ, is an auxiliary organ of the sense of smell in some vertebrates such as the white-tailed deer, all of which are tetrapods. It is in the vomer bone, between the nose and the mouth. Sensory neurons within the organ detect different chemical compounds, usually large molecules. Snakes use it to smell prey, sticking out their tongue and attracting particles to the opening of the organ on the palate. Some mammals use a characteristic facial movement called flehmen reflex to send compounds to this organ, while in other mammals the same organ contracts and pumps to attract the compounds. Most animals with a vomeronasal organ use it for the detection of pheromones even though some pheromones are detected by the organ of smell. The vomeronasal organ seems to detect other compounds besides pheromones. Among mammals, sex pheromones are almost always detected by the vomeronasal organ. Odors produced by an individual and detected by another of the same species are called pheromones if the process involves real communication and benefits both individuals.

The neurons that receive the vomeronasal organ have axons that leave the capsule of the vomeronasal organ in groups and extend dorsally passing below the olfactory mucosa. The axons carry these electrical signals to the accessory olfactory bulb. This processes the information and is located dorsal and medial to the central olfactory bulb, which processes the olfactory information. The central olfactory bulbs receive the information of groups of olfactory axons that come from the olfactory receptor neurons in the olfactory mucosa within the nasal cavity. The information of the vomeronasal organ of the accessory olfactory bulb and the olfactory information of the central olfactory bulb is transported separately by second order axons to the amygdala. From there, the vomeronasal organ system is projected directly into the preoptic area and the hypothalamus; the areas that are involved in reproductive behavior. The projections of the central olfactory system include the hypothalamus, the thalamus and the cortex.

Body development and nutritional status

Physical and physiological indices taken in the deer, for the determination of their nutritional status, have been developed only under controlled conditions and usually with balanced diets. Such indexes have rarely been used to assess nutritional stress in wild populations of deer, and on very few occasions have influenced their management policies. However, for a specific index of the nutritional stress condition, five blood components of deer:

1. Nitrogen in urea used as an indicator of protein consumption and protein deficiency. This parameter can be affected by the ability of the deer to recycle urea via saliva or by direct infusion into the rumen.
2. Creatinine, which is a non-protein nitrogenous compound that occurs due to the catabolism of muscle tissue. Its increase may be associated with muscle necrosis, atrophy and malnutrition.
3. The total protein provides a good reference of protein consumption, liver function and immunological status.
4. Free fatty acids are produced in response to the catabolism of adipose tissue and are usually inverse to the amount of energy in the diet.
5. The serum concentration of the thyroid hormone triiodothyronine (T3) has been particularly valued as an indicator of the nutritional stress of deer.

Of all the indices studied, the serum concentration of the thyroid hormone triiodothyronine (T3) has been particularly valued as an indicator of the nutritional stress of deer, specifically of low concentrations of energy in the diet or body of the animal. A combination of the determination of T3 concentration in the serum and a ratio of the weight and width of the chest has been recommended as an indicator of energy status in deer. The serum concentration of another thyroid hormone, thyroxine (T4), has been associated with protein intake in the diet and with the concentration of energy. Likewise, it has been reported that the T3 hormone intervenes in the growth of the antlers in the deer; this is shown by a study who demonstrated that the increase of the hormone is associated with the development and mineralization of the antlers of white-tailed deer; however, the levels of the hormone T4 and the enzyme alkaline phosphatase were not correlated with the growth and mineralization of the antlers.

Chapter 2

Digestive physiology

> **Summary.** The rumen of the deer provides an appropriate environment, with enough food for growth and reproduction of the microorganisms. Anaerobic conditions in the rumen favor the development of bacterial species, among which they can digest the walls of plant cells (cellulose) to produce simple sugars (glucose). The microorganisms ferment glucose to obtain the energy to grow and produce volatile fatty acids (VFA) as final fermentation products. The VFA cross the walls of the rumen and serve as sources of energy for the ruminant. As rumen microorganisms grow, they produce amino acids, protein precursors. Bacteria can use ammonia or urea as sources of non-protein nitrogen to produce amino acids. However, the bacterial proteins produced in the rumen are digested in the small intestine and constitute the main source of amino acids for the animal. The bacteria constitute half of the biomass in the normal rumen and are responsible for metabolic activity. The fungi constitute up to 8% of the intra-ruminal biomass and are located in the intake of slow movement avoiding its rapid washing; in addition, they intervene in the digestion of low quality forages. Protozoa are the most notable organisms in the rumen, form a large proportion of the biomass, between 20 to 40%, but their contribution is lower due to the great retention and the lower metabolic activity. Its generation time is large and the survival in the rumen depends on the strategies that reduce washing. Fungi release a more soluble cellulose complex than bacteria and attack rough particles to which they ferment more rapidly than bacteria. Highly ground or concentrated foods have less fungi. The fungi produce VFA, gases and traces of ethanol and lactate. The constant absorption of the metabolites and the ions produced in the deer rumen contributes greatly to the maintenance of conditions necessary for the existence of an adequate ruminal flora.

Introduction

Small herbivores select a better-quality diet because of their relatively higher nutritional requirements. This is related to a distinctive feature of small ruminants classified as browsers such as white-tailed deer that have parts of the mouth that allow them to be more selective with plants and parts of plants, than ruminants classified as grazers such as bovines. The forages selected by the deer pass quickly through their digestive tract (DT). The digestive physiology of the deer has particular aspects. The degradation of the food is mainly done through fermentative and non-enzymatic digestion. In addition, the fermentative processes are carried out by different types of microorganisms (protozoa, fungi and bacteria) that the deer harbors in its DT. In addition, for there to be a good nutritional ecology there must be a favorable ruminal environment, developing a symbiosis between microorganisms and the deer. The fermentation process is carried out mainly in the first two parts of the DT by the ruminal microorganisms and the physical and chemical medium that surrounds them. The main products of ruminal fermentative processes are volatile fatty acids (VFA).

Structure and functionality of digestive system of deer

Mouth.- It is the entrance of the digestive system. It is a cavity between the maxillary and palatal bones, elongated according to the axis of the head, and with two openings, one anterior and one posterior.

Esophagus.- It is a long muscle-membranous tube, placed between the pharynx and the stomach, which is responsible for conducting food during swallowing. It leaves the lower part of the pharynx and goes from top to bottom and from front to back, behind the larynx and trachea at the lower edge of the neck.

Rumen, reticulum, omasum and abomasum.- The white-tailed deer is a ruminant and, like all ruminants, has a stomach divided into four compartments: rumen, reticulum, amaso and abomasum (the latter is the true stomach). These compartments develop from the embryonic stomach and are relatively small in the newborn deer, where the rumen and reticulum together are barely half of the abomasum. The considerable growth of the rumen develops during the first months of life. life, but the main stimulus of its development is the ingestion of solids. When the four compartments reach their relatively permanent size, which occurs after one year, the rumen represents 80% of the total volume of the stomach.

The four organs fill almost three quarters of the abdominal cavity, occupying virtually the entire left part and extending significantly to the right. The reticulum is located adjacent to the rumen only separated by a fold called esophageal groove. The rumen is the largest of the four stomachs; It is composed of four bags called: dorsal, ventral, dorsal and caudo-ventral. The reticulum-rumen is considered a cranial-ventral sac of the rumen because the intake flows freely between these two organs. The reticulum connects with the omasum through a small tunnel. The abomasum is the true stomach of the ruminant. Histologically it is very similar to the stomach of non-ruminants.

The interior of the rumen, reticulum and omasum are covered exclusively with a stratified squamous epithelium, very similar to that observed in the esophagus. However, each of these organs structurally have a very different mucosa. The inner surface of the rumen forms numerous papillae that are different in shape and size. The reticular epithelium forms a series of folds that constitute polygonal cells, which give the appearance of a honeycomb. The inner lower part of these polygons is formed by many small papillae. In white-tailed deer, polygonal cells are shallower than those of a bovine. The interior of the abomasum is composed of longitudinal folds or leaves that look like the pages of a book. The abomasum folds represent one third of the total surface of the front stomach of the ruminant.

Small intestine.- It is the narrowest part of the intestine, its caliber is uniform and its length varies, but it is always many meters. It is cylindrical, coiled in spiral, and has two curvatures called large and small curvature, this is what serves for the insertion of the mesentery. Presents three equal parts or portions: duodenum, jejunum and ileum, which communicates with the blind. The duodenum is the first portion of the small intestine and is where the biliary and pancreatic digestive secretions are poured, which, in conjunction with the gastric and intestinal juices, unfold the nutrients of the intake in their absorbable forms.

Large intestine.- It continues to the small intestine, from which they are easily distinguished by their caliber, which is many times greater, and by a series of strangulations and dilatations or bulges, which give it a special aspect. It begins in a dilatation or very large reservoir, called blind, which continues with the part called colon, which consists of two sections: the collapsed colon and the floating colon, ending with the rectum. The main function of the large intestine is the absorption of water. This is how the total dry matter of the intestinal content increases from 7% in the near sector of the large intestine to 15 to 18% in the feces.

Rectum.- It is the continuation of the floating colon. It is given the name of rectum, by its provision in straight direction, from front to back. It ends in the anus which is the posterior opening of the digestive tract, which makes it communicate with the outside. The rectum serves as a reservoir bag, where excrement is stored in the range of bowel movements. Its structure is a fleshy, thick layer, which is pink in color, and has numerous longitudinal and transverse folds. It lacks a serous layer, except in the part before the entrance of the bassinet.

Anus.- It is the posterior opening of the digestive tract. It is located below the tail. In its outline it resembles the opening of a bag that is closed by means of a sliding knot, forming a runner, all the more protruding while the animal is younger and more vigorous. Its structure is mucous on its internal side, which is transitional between the skin and the true mucosa, then muscular, in the form of a fleshy, reddish, called the sphincter of the anus: it is the layer that keeps the anus closed at intervals of defecations, and externally a layer of fine skin without hair that is unctuous and soft, due to the large number of sebaceous glands it contains.

Fermentation, microbiology and ruminal ecology

The fermentation in the rumen of the deer is carried out by a rich and dense range of microbes. Each milliliter of ruminal fluid contains approximately 10 to 50 billion bacteria, one million protozoa and in lesser, but more variable, fungi and yeasts whose function of the latter is not well defined. The environmental conditions of the rumen and large intestine are anaerobic and, as might be expected, almost all microorganisms are anaerobic or facultative anaerobic. In a complex food web, fermentative microorganisms interact and feed each other; that is, the waste products of some microbes serve as food for other species.

Bacteria.- The different genera of fermentative bacteria make up a chamber with ample digestive capacities. These organisms are classified according to the substrate they prefer or the final products they produce. Even when there is some specialization, many bacteria use the same substrates. However, there is considerable overlap between species of bacteria by substrates or their products.

The bacteria of the rumen are those that perform several of the vital functions for the development of the deer. Fibers and other insoluble plant polymers that cannot be degraded by the enzymes of the deer are fermented to AGV, mainly acetic, propionic and butyric, and to gases such as CO_2 and CH_4. The VFA pass through the walls of the rumen and pass into the blood, then oxidized in the liver and become the largest source of energy for the cells. Fermentation is coupled to microbial growth and biomass proteins are the main source of nitrogen for deer. In addition to the digestive functions, the rumen microorganisms synthesize amino acids and vitamins, mainly of the B complex, being the main source of those essential nutrients for the deer. Also some bacteria degrade toxic components of the diet such as amino acids mimosina and its derivatives, plant phenols such as coumarin (1,2-benzopirona), canavanina, analogous to arginine, which inhibits some bacteria of the rumen, but is hydrolyzed by other . The rumen includes between 1010 and 1011 bacteria/ml, and more than 75% is associated with food particles. The general density does not vary with the diet, but the number of different species is associated with the availability of the substrate for fermentation.

The rumen bacteria of the deer are predominantly strict anaerobes, but also coexist with facultative anaerobes, attached to the walls of the rumen, these use the O_2 that comes from the bloodstream and are very important in the functions of the rumen being the most important ones that ferment cellulose Many are highly specialized, have numerous nutritional requirements that must be provided by the system. Others, on the other hand, use few sources of energy and another group are more inconsistent in the energy requirement.

When two or more microorganisms combine their metabolic capacities to degrade a substance that can not be catabolized in an individual way by either of them, the concept of syntropy is reached, in the rumen there are syntrophic groups related to the degradation of fibers, including, for example to the cellulolytics, hemicellulolytics and the microorganisms that happen them, like the methanogenic bacteria. The most competitive ones present adhesion to the substrate and energy storage within the cell.

Cellulolytic bacteria.- The degradation of cellulose is the main function of the rumen. The active bacteria in the rumen adhere to plant fragments and secrete their hydrolytic enzymes that release soluble oligosaccharides, mainly cellobiose, used by the cellulolytic microflora and by other microorganisms that do not degrade cellulose (Table 2.1). If it is not hydrolyzed, an inhibition of another group of bacteria can occur when the substrate they require is not present. It is possible that glucose, another product of cellulolysis, can inhibit the activity of some enzymes as well. Many of the cellulolytic species can also degrade the fraction wrongly called hemicellulose.

Table 2.1. Utilization of carbohydrates for ruminal bacteria

Species of bacteria	Polysaccharides	Mono- y disaccharides
Celluloliticss:		
Fibrobacter succinogenes	cellulose, celodextrinas	G, C
Ruminococcus flavefaciens	cellulose, xylene, pectin	C
Ruminococcus albus	cellulose, xylene, celodextrinas	G, C, X, A
Celulolíticas secundarias:		
Butyrivibrio fibrisolvens	cellulose, xylene, dextrin, pectin,	G, Ga, Mn, F, M, X, L, C
Clostridium longisporum	cellulose	G, Ga, F, C, M, L, S
Clostridium locheadii	cellulose, dextrin	G, M, S
Non cellulolitics:		
Prevotella ruminicola	pectin, starch, dextrin,	G, Ga, F, L, C, X, A, R, M
Selenomonas ruminantium	starch, dextrin, celodextrinas	G, Ga, F, X, A, C, M, L, S
Streptococcus bovis	starch, celodextrinas	G, Ga, F, Mn, C, M, L, S
Succinivibrio dextrinosolvens	dextrin, pectin	G, Ga, Mn, X, M, A, F, S

A: arabinose, C: cellobiose, F: fructose, G: glucose, Ga: galactose, L: lactose, M: maltose, Mn: mannose, R: rhamnose, S: sucrose y X: xylose.

Amylocytic bacteria.- Most of them do not use cellulose. The amylolitic enzymes are widely distributed among bacteria and are the ones that ensure the conversion of starchy materials, such as cereal grains, into AGV. With the presence of ammonia, the process is more efficient. The main source of energy is obtained through the fermentation of carbohydrates. Of which the microorganisms obtain energy, with liberation of AGV, hydrogen, carbon dioxide, water, methane. The most important VFA are acetic acid, propionic acid and butyric acid. The energy use is greater when propionic acid is produced than when acetic acid is produced, since in the latter, H_2 and CH_4 are released, which are forms of dissipated energy. The animal takes advantage of the VFA as the main source of energy through the absorption of the same, through the ruminal wall. It should be mentioned that the effect of rumen pH is of great importance, given that it infers in the different chemical processes, population levels, interactions, regulatory systems, among others

Protozoa.- Its main function of protozoa is to ingest particles the size of bacteria, such as starch, fibers, chloroplasts. Most of the components are Ciliata, the most complex unicellular organisms. Its biomass is like that of bacteria, but they can exceed it more than 3 times according to the diet, or even disappear. The different species vary in size, between 25 and 250 microns, grouped into genera of the subclass Entodiniomorphes and the subclass Holotriches, which differ in their morphology and metabolism. The species present vary with the animal species, the locality, the diet and its cellulolytic activity. Generation times range from 0.5 to 2 days. The slower ones can disappear with the fluids of the rumen, several remain attached to fragments of food, so they are more retained than bacteria and a large part can be lysed in the rumen of the deer.

Ciliates differ from bacteria in several aspects:

1. They are very mobile and invade newly ingested foods as fast as bacteria despite being in smaller numbers.
2. Additional carbohydrates can be stored in the form of insoluble polymers, amylopectin.
3. They are more easily destroyed by acidity, Holotriches are more sensitive than Entodinomorphes.

4. They cannot synthesize amino acids from simple compounds of nitrogen and depend on bacteria, using amino acids after phagocytosing (1% of bacteria are phagocytized every minute).
5. They are responsible, in large part, to produce ammonium in the rumen.
6. The ciliates are not essential for the fermentation processes, but they help to make them more efficient.

Cellulolytic ciliate.- Few genera of Epidinium are involved in the fragmentation of plant remains. These secrete enzymes that cause the separation of cells and the fragmentation of the material. More than half of the cellulolytic activity of the rumen is associated with ciliates. The greatest activity occurs when the enzyme is released after cell lysis that occurs due to exposure to O2 in rumination or hypotonia caused after water ingestion.

Amylolithic ciliates.- All Entodiniomorphes use starch whose excess they store as amylopectin. But one of the two genus Holotriches cannot use starch. Most prefer soluble sugars and move quickly towards them. Other sources of energy: the ciliates are responsible for 30-40% of lipolysis. Increase the content of saturated fatty acids. 75% of microbial lipids are normally associated with ciliates. They are not very important in the degradation of proteins in the diet, they use those of phagocytized bacteria. The rumen is a complex ecosystem, which is found in a dynamic way, influenced by the entrance from the outside of the food, water, other microorganisms, etc., the exit of the materials to the intestine, and by the complex interactions that occur within East. It must be borne in mind that it functions as a fermentation chamber where almost anaerobic conditions prevail (there is approximately 0.6% oxygen), with reducing conditions, slightly acidic pH, and temperature around 39° C.

Since the discovery of ciliated protozoa in ruminants, many works have been carried out on their morphology, their functions and their relations with the host. Secondary compounds (phenols, tannins), present in many browse plants, could act as elimination protozoa agents to control protozoan populations in ruminants. Removal does not necessarily refer to the total elimination of the protozoa of the ruminal ecosystem, rather elimination is a sustainable factor for metabolizable energy in ruminants, since it is a factor that increases energy production. Comparative studies have been carried out on animals with controlled fauna to evaluate the specific function of a population or species. The conclusion of these studies is that the effects of elimination of protozoa depend on the method of used and the diet consumed by the ruminant. However, the presence of protozoa can alter numerous factors such as: cellulolytic activity of ruminal bacteria; physiological factors; rumen environmental conditions (pH, AGV, N-NH3) and retention time of the substrate in the rumen. The large biomass of protozoa that exists in the rumen is 40 to 80% of the microbial mass, and its ability to digest the largest number of food components suggests that these have an important role in the ruminal fermentation of the deer.

On the other hand, a positive effect of the cactus supply in NH3-N concentrations was found to be accompanied by an increase in the total protozoa count in the rumen fluid ($P < 0.001$). The average protozoa count changed from 3.5×10^4 ml-1 to 13.0 17.7 and 13.1×10^4 ml-1 with diets supplemented with 0, 300, 450 and 600 g of MS, respectively, of prickly pear cactus. The high protozoal count observed in animals supplemented with prickly pear was associated with high concentrations of NH3-N in the rumen. It is assumed that protozoa contribute to the digestion of dietary protein and the production of ammonia.

The food that the deer consumes and that must be digested by the microorganisms of the microbial ecosystem of the rumen, is physically of a much larger size than the microorganisms themselves. Therefore, microorganisms must solubilize foods by secreting enzymes to chemically digest the food into smaller soluble components. There are different enzymes that are manufactured intracellularly by microorganisms to be excreted, in different ways, into the rumen environment. These enzymes include cellulases, hemicelluloses, pectinases, amylases, proteases and lipases. There are multiple enzymes within each class that act concurrently by hydrolyzing the substrates into soluble and absorbable forms.

An important peculiarity of some protozoa is their ability to regulate glycogen synthesis; When soluble carbohydrates are in abundance, they continue to store glycogen until it is completely filled. Another characteristic of protozoa is that many species consume bacteria, this fact is thought to play an important role in the control of bacterial growth.

The distribution of the different microbial species varies with the type of diet consumed by the deer. The presence of some of them, apparently reflect the availability of their substrates; For example, populations of cellulolytic microbes decrease when the diet of the deer is rich in grains.

Fungi.- Flagellated fungi possess mobile zoospores and colonize damaged regions of plant tissues within two hours of ingestion, in response to soluble materials. At 22 hours more than 30% of the larger particles are invaded by rhizoids. Its main role is to facilitate the disappearance of the cell wall of the plant cell. Species of 4 genera have been identified: Neocallimastix, Caecomyces (formally Sphaeromona), Pyromyces (formally Phyromonas) and Orpinomyces. Its life cycle involves a fruiting body (sporangium) originated from a mobile zoospore that adheres to the fibers and develops sporangia and rhizoidal filaments, which penetrate the lignocellulosic matrix, where the enzymes act. Fungi release a more soluble cellulose complex than bacteria and attack rough particles to which they ferment more rapidly than bacteria. Highly ground or concentrated foods have less fungi. The fungi produce VFA, gases and traces of ethanol and lactate.

The environmental conditions of the deer rumen have profound effects on the microbial flora and fauna. Ruminal fluid normally has a pH between 6 to 7 but can decrease if the deer consumes large amounts of soluble carbohydrates. If the pH decreases around 5.5, protozoa populations are markedly reduced due to their intolerance to acidic pH. If the pH drops further, which could occur with an over consumption of grains, many species of microorganisms could be destroyed with serious consequences for the deer.

Ruminal bacteriophages.- Within the microbiology of the reticulum-rumen complex, viruses have been identified and isolated. It has been found in concentrations of 1010/ml, identifying from 26 to 40 different forms, have been classified into three families Myoviridae, Siphoviridae and Podoviridae. The bacteriophages do not contribute to the fermentative process or the degradation of substrates, nor do they have respiratory activity, and they are obligate pathogens of the bacteria lysing them and recycling protein by an inefficient use of the food, although authors maintain that Bacteriophages participate in the recycling of limiting nutrients, helping to maintain the ruminal ecosystem.

The fermentation chamber

The deer rumen is a perfect fermentation chamber that provides the convenient anaerobic environment for the continuous cultivation of the microbial population. The correct assiduity of conditions in the rumen is achieved in the following way:

1. Frequent food intake by the deer provides a regular substrate supplement for the microorganisms.
2. The soluble products of the microbial activity are quickly absorbed by the ruminal wall, and therefore do not accumulate or even inhibit the enzymatic action.
3. The osmolarity of the rumen close to 300 mosm.
4. Maintaining an oxide-reduction potential.
5. The rumen temperature to be from 39 to 40 °C.
6. The volume of the ruminal content is regulated by the passage of the liquid material at intervals towards the omasum through the reticulum-omasum orifice. The small food particles and a portion of the microbial population are removed from the rumen in this way.
7. Ruminants secrete large amounts of saliva that is rich in bicarbonate and other ions. Saliva is the most important factor for maintaining liquid volume and fixes the pH status of the ionic composition in the rumen.

The ruminants like the deer were developed to consume and subsist with small seeds, fruits, grasses and mainly of the foliage of bushes (browse). They are very selective in their diet, tend to have simple rumen and consume and chew frequently. The reason for this behavior is that the deer consumes a diet of very high quality and, therefore, with high production rates of volatile fatty acids. The fact of consuming constantly, allows the deer to avoid the spikes of AGV production that could be pathological. Animals such as deer tend to consume complete seeds, fruits and herbs, the lack of chewing of these foods, allows cell surfaces are not broken, with the above, limit and regulate the speed of fermentation and production of AGV. It is known that approximately 50% of the organic carbon in the earth is bound to the cellulose molecule. What represents a huge source of energy, even when the cells of vertebrates do not produce the amount of cellulases necessary to break all this abundant material. However, many microbes secrete cellulases that allow them to use cellulose from diet and other plant materials. The polysaccharide cellulose as well as other carbohydrates from the diets of ruminants, during ruminal fermentation, are converted into volatile fatty acids, which are eventually transported, through the ruminal wall, into the bloodstream.

Dynamics of cranial digestion of deer

Once the food, water and saliva enter the mouth of the deer, they are sent, via the esophagus, to the reticulum-rumen. Heavy materials (grain, rocks, nails) remain in the reticle, while light materials (grass and hay) enter the rumen. In addition to all this mixture of materials, there are voluminous amounts of gases produced during fermentation. All these foods are retained in the reticulum-rumen until they reach a fine consistency. The saliva that is produced in abundant quantities serves as a lubricant as a supplier of fluids for rumination and fermentation and alkaline buffer as mentioned above.

The adhesion of the microorganisms to the cell walls is a process prior to the digestion of carbohydrates. Ruminal microorganisms are classified by into three subpopulations, according to their interaction with food particles: 1) those associated with ruminal fluid; 2) those weakly associated with the particles; and 3) those firmly attached to said particles. Within the first group are the microorganisms that use the soluble nutrients of the ruminal fluid, in addition to those that are detached from the food particles. As free organisms, they have little role in the digestion of solid particles, but many of them have adhesion capacity.

The remaining two groups make up between 70 and 80% of the ruminal population. Weakly adhering organisms can be detached by washing the particles and are generally adhered to by nonspecific mechanisms involving physicochemical attractions or interact with other microorganisms attached to the particles. The velocity of solid material passing through the rumen is quite slow and depends on its size and density, although it is faster in deer than in bovines. Water flows fast through the rumen and, apparently, its presence is important to push the particles into the lower tract. Due to the fermentation, the food is reduced to more and more small particles, as well as the constant proliferation of ruminal microorganisms. Ruminal contractions, at the same time, constantly push the light solids back into the rumen. The smaller and denser material tends to be pushed towards the reticulum and the cranial sac of the rumen, from which the fluid loaded with microbes is sent, through the reticulum-omasum orifice, towards the omasum.

The function of the omasum in the stomach of the deer is not fully understood. It probably works to absorb residual volatile fatty acids and bicarbonate. Apparently, its main function is to absorb water and some particles that get stuck in the omasum folds. However, the constant contractions of the omasum allow the solid material to move towards the abomasum. The abomasum is the true stomach of the deer, which secretes acid and has functions very similar to those of the stomach of the norumiants. It is specialized to process large amounts of bacteria. Contrary to the stomach of the norumiants; In addition, the abomasum secretes lysozyme, an enzyme that efficiently degrades the cell walls of bacteria. The secretion of hydrochloric acid (HCl) towards the abomasum, to form the gastric juice, is regulated hormonally and by the presence of peptides in the ingestion. Once the HCl enters the abomasum, the pH can drop to values of 1.5 to 3.0. In these acidic conditions, the protozoa that have passed to the abomasum from the rumen disintegrate and some of the bacteria die.

The process described above applies to adult ruminants. However, the newborn ruminant behaves like norumiante. The front stomach is fully formed, but it is not fully developed. If the milk enters the rumen at that age, it rots, instead of being fermented. To avoid such problem, in young ruminants, the action of breastfeeding produces a closed reflex action of the collapsible muscles that form a channel from the esophageal orifice towards the omasum forming the groove or esophageal groove, sending the milk directly to the abomasum, where it is curdled by the action of the enzyme renin and subsequently digested enzymatically in the duodenum.

Digestion in the small intestine

In the digestion, in charge of digestive enzymes, the prevailing pH conditions in the intestine play an important role. In the case of deer, the neutralization is slower, probably due to the large amounts of hydrochloric acid secreted with the gastric juice, as well as the lower alkalinity and lower bicarbonate content of the secretions and pancreatic. The cecum is located at the junction of the small intestine with the large intestine, which is a lateral sac. This compartment is connected to the digestive tract by a single opening. Both the pH and anaerobic conditions in this cavity give rise to a new process of microbial fermentation of those nutrients that until now have not been digested or absorbed by the ruminant. However, this fermentation is not of fundamental importance for the deer, both because of its small volume and because of the low rate of absorption in the large intestine of the compounds resulting from this process.

    The amount of nutrients that reach the small intestine in the deer is relatively much lower than that which enters the diet due to the great fermentation that takes place in the rumen. The animals that consume diets based on forage, apparently only 5 to 8% of the rapidly digestible carbohydrates, escape ruminal digestion. Other nutrients from the diet that reach the small intestine are the lipids of plants and variable proportions of cell wall carbohydrates and proteins that are not degraded by ruminal microbes. In addition to these diet contributors, substantial amounts of microbial cells of ruminal origin enter the small intestine.

    Digestion in these sites is intervened by the presence of acid in the abomasum and the existence of more than 40 digestive enzymes. Enzymatic digestion is initiated by secretions of the abomasum mucosa, but occurs mainly in the small intestine where it is mediated by pancreatic and intestinal enzymes. It is likely that there are certain limitations to carry out the digestion of microbial nutrients and diet that reach the small intestine when the deer are consuming diets with high content of forage, another that does not correspond to the residues of cell walls fermented. A great digestion of these nutrients can be assisted by the continuous digestive nature in the deer. Grazing animals invest a large proportion of their time in rumination. These activities, accompanied by the large amount of intake in the reticulum-rumen, result in a flow of intake to the postruminal tract in a continuous manner. This, in turn, stimulates the continuous release of digestive secretions into the abomasum and small intestine.

Fermentation in the caecum and colon

Ruminants gain some benefit during secondary fermentation in the cecum and in the proximal colon. These compartments are characterized by the presence of an active microbial population. When the diets are based on forage, the main substrates that enter the caecum and colon are the indigestible materials composed of the cell wall of the plants and indigestible components of the bacterial cells. These substrates are more refractory than those that enter other segments of the digestive tract, although the susceptibility of cell wall residues to microbial attack can be improved once they are exposed to abomasum and intestinal digestion and by reductions in the size of food particle. Digestion in the cecum is aided strictly by microbial activity; there are no digestive enzymes associated with the mucosa of the caecum and colon.

Digestion in the cecum is less intensive than in other sites of the digestive tract because the substrates that reach this site are more refractory and because the retention time is shorter than in the rumen (7 to 8 hours). The counts of viable bacteria in the content of the cecum are in a range of 107 to 109 g^{-1} of digested material. The species of important bacteria in the caecum and colon are identical to those in the reticulum-rumen. Although the relative numbers vary. Protozoa are absent in the blind of ruminants. The physical-chemical conditions of the cecum are like those of the reticulum-rumen; the pH is from neutral to slightly acidic (6.0-7.7) and the total AGV concentrations are in a range of 60 to 90%, which are typically observed in the rumen of deer.

Reticulum-rumen motility

An orderly pattern of ruminal movements begins when the ruminant is born and, with the exception of small recess periods, they persist throughout the life of the deer. These movements serve to mix the intake, help the belch to expel the gas, drive fluids and fermented products towards the omasum. If the motility is suppressed for a significant period, the result is a paralysis of the rumen.

A cycle of contractions in the front stomach of the deer occurs one to three times per minute. The fastest frequency occurs during feeding and the lowest frequency when resting. Two types of contractions can be identified:

1. Primary contractions that originate in the reticulum and go caudally to the rumen. This process involves a contraction wave followed by a relaxation wave, as one part of the rumen contracts, another sac dilates.
2. Secondary contractions that occur only in certain parts of the rumen and are usually associated with belching.

The ruminal motility is controlled by the nervous system of the deer. The front stomach possesses a rich enteric enervation, although the contractions are coordinated by the central nervous system. The motility centers in the brain control the speed and duration of contraction through the nerve vagus. There are several afferent branches of the rumen to the motility centers which allow to narrow the receptors and chemical receptors in the rumen to modulate the motility. Conditions within the rumen can significantly affect motility. Yes, for example, ruminal content is very acidic; what happens when there is a high grain content. Likewise, the type of diet influences the motility: if for some reason the deer consumed a diet with a high straw level, their stomach would have more contractions than if they had a diet based on grains.

Rumination and belching

Rumination is defined as those mechanisms that involve the re-insalivation, re-masticating and swallowing of the rumen content in the ruminant. Rumination for deer is important because the rumination carried out by the deer, consists of regurgitating the intake found in the reticulum, to carry out a re-stratification and, later, to swallow the remasticated material. Rumination provides an effective mechanical breakdown of the forage:

1. contributing to the degradation of the particle size of the food
2. increasing the specific gravity of the forages

3. breaking the impermeable cover of the plant tissue, increasing the surface area of the forage for the adhesion and ruminal digestion that the microbes carry out and
4. increasing the surface area of the substrates for fermentation to take place.

Regurgitation starts with a reticular contraction, different from the primary contraction. This contraction, in conjunction with the relaxation of the distal sphincter of the esophagus, allows the bolus of intake to enter the esophagus. The bolus is driven towards the mouth by peristaltic movements in reverse. The fluid that contains the bolus, which is squeezed by the tongue of the deer and the bolus, which is remasticated, is subsequently swallowed. Rumination occurs predominantly when the animal is resting and is not consuming. In wild ruminants, such as white-tailed deer, rumination takes place under the shelter of vegetation, hidden from predators.

The maximum average size of food particles that pass from the rumen to the omasum is <1.0 mm for almost all ruminant species. However, body size is the main factor that determines the efficiency of rumination. Larger animals, such as cattle, are more efficient in rumination. Small ruminants such as deer do not chew large amounts of plant cell walls, so they must select high-quality plant material.

The fermentation in the rumen generates huge and dangerous amounts of gases. About 5 liters per hour in the white-tailed deer. Belching is how the ruminant constantly removes the gases produced by fermentation. As previously mentioned, each belching is associated with a secondary rumen contraction. The belched gas travels towards the esophagus at a speed of 160 to 225 seconds, but before being expelled, most of the gas is inhaled by the lungs, then it is exhaled. Any factor that interferes with belching endangers the life of the ruminant because the expansion of the rumen quickly blocks breathing. The animals that suffer ruminal tympanism die by asphyxia. Tympanism is rare in white-tailed deer, unless it consumes a toxic plant that causes tympanism.

Absorption of nutrients

The absorption of metabolites and ions in the rumen of the deer is a factor of the greatest importance for the maintenance of reasonably constant conditions of pH, ionic composition, etc., which are necessary for the existence of a healthy rumen flora. The wall of the rumen, through which absorption is performed, is covered by an epithelium stratified with papillae, supported by a layer of smooth muscle fibers. The epithelium is not glandular and has no secretory function. The most important metabolites that are absorbed in the rumen are:

1. Volatile fatty acids (VFA) that are produced in the rumen in large quantities during ruminal fermentation are rapidly absorbed. Virtually all the amount of the acids, acetic, propionic and butyric formed in the rumen are absorbed through the rumen epithelium, from which they pass to the portal vein that subsequently leads them to the liver. The continuous removal of AGV from the rumen is important, not only for distribution, but to prevent excess and damage that could be produced by the decrease in the pH of the ruminal fluid.
2. Ammonia, which is a final product of the degradation of urea, amino acids, peptides and proteins in the rumen, is also rapidly absorbed.
3. Inorganic ions, mainly potassium and calcium chlorides and phosphates, and water that passes rapidly through the rumen epithelium by the osmotic changes in ruminal content, are efficiently absorbed.
4. Glucose and lactic acid, the presence of these metabolites in the rumen is due to there being an excessive accumulation of them so they can be absorbed in the rumen.
5. Gases, carbon dioxide and methane that are products of fermentation and most of these are eliminated by the lungs in the belching; It has been shown that a small part is absorbed in the rumen.

Chapter 3

Water

> **Summary.-** Water is the most critical nutrient for deer because it requires regular consumption since an adult contains approximately 50 to 66% of their body mass and up to 90% of newborns and more than 99% of the molecules of their organism. All the biochemical reactions that take place in the deer organism require water. Many of the biological functions of water are dependent on the property it must act as a solvent for a wide variety of compounds, and many of these compounds are easily ionized in water. In addition, water serves as a means of transporting viscous substances and digested semi-solid substances in the digestive system, of various solutes found in the blood of deer, in tissue fluids and in cells and in excretions such as urine and water. sweat. Also, water is part of many reactions. The feces of the deer, in comparison with those of the bovines, contain less humidity, reason why, keeping the proportions, in the intestines of the deer more water is absorbed than in those of the bovines. Forage plants usually contain abundant amounts of water from 45 to 65% in forage shrubs and 70 to 90% in herbs, fruits and flowers. The deer loses water through: 1) urine; 2) feces and 3) insensitively, which is water that is lost through vaporization in the lungs and dissipation through the skin and sweating through the sweat glands, when the temperature is warm. Losses through urine and feces occur at intervals, but losses through the lungs and skin are constantly present. In the case of a more severe restriction, a rapid loss of weight is shown as the deer is dehydrated.

Introduction

Chemically pure water is the combination of hydrogen with oxygen. In the natural state, it is clear, without color, or smell. Water is part of the diet of the animal, and after oxygen, it is the most important and indispensable component for life on earth. Water constitutes the greatest weight of animals and vegetables. The lack of water can cause death quickly, more than the lack of any other element. In its liquid or solid form, it covers more than 70% of the planet. 69% of the total world water is used for agriculture, 23% for industry and 8% for domestic needs. The deer use water for their nutrition and growth, and they obtain it from three sources: the one contained in the food, the one that is produced during the process of assimilation of the same, and the drinking water. From the physical point of view, water acts on the deer as a buffer between its own temperature and the environment. From the nutritional point of view, it behaves like a universal solvent. The water favors the softening and fermentation of food, allowing its assimilation and the excretion of urine and feces. Water is the main cellular constituent, forming part of more than half the weight of the deer.

Physical properties and functions

Water is the most critical nutrient for deer because it requires regular consumption. An adult contains approximately 50 to 66% of their body mass and up to 90% of newborns and more than 99% of the molecules in the body (the above is possible because the water molecules are smaller than the others). Apparently, the deer can survive for about a month with little or almost no food, although it can die in three days if it is devoid of water. Deer can lose weight and stop their feed intake even with a moderate water restriction.

The hydrogen bond between H and O in the water molecule has special characteristics and is the most related of all the links that exist in nature; however, it is the most unstable. This particularity makes water have special physical properties that make it essential for the life of living beings such as deer because water has two basic functions for all terrestrial animals: 1) it is the main component of body metabolism and 2) is the main factor in the control of body temperature.

All the biochemical reactions that take place in the deer organism require water. Many of the biological functions of water are dependent on the property it must act as a solvent for a wide variety of compounds, and many of these compounds are easily ionized in water. In addition, water serves as a means of transporting viscous substances and digested semi-solid substances in the digestive system, of various solutes found in the blood of deer, in tissue fluids and in cells and in excretions such as urine and water. sweat. Also, water is part of many reactions. In hydrolysis, water is a substrate of the reaction and in oxidation, water is a product of the. Water is very important in the control of body temperature because it has a high specific heat, high thermal conductivity and high latent heat of vaporization, which are properties that allow the deer to accumulate heat, can easily transmit it and can lose by evaporation. Also important are the lubrication of the joints and the damping of these, of the organs that are inside the deer and of its central nervous system by means of the cerebrospinal fluid. In addition, water provides the basic means for the conduction of sound in the middle ear of the deer and contributes in the transmission of its other special senses.

Water absorption

Water is absorbed very quickly in most sections of the digestive system of deer. In the rumen and omasum of the deer, there is usually a net absorption of water. The same applies to the duodenum where the pancreatic, biliary and intestinal glandular fluids produce a net absorption of water. In all species there is a net absorption in the ileum, jejunum, caecum and large intestine, but the amount that is absorbed varies considerably, since it depends on the moisture in the feces. The feces of the deer, in comparison with those of the bovines, contain less humidity, reason why, keeping the proportions, in the intestines of the deer more water is absorbed than in cattle.

This characteristic of deer allows them to retain more water and enables them to survive for prolonged periods in regions where water is scarce. The type of diet consumed by the deer also influences the absorption of water in their digestive treatment. If the deer consumes grasses and sprouts of pastures, it will absorb more water, on the other hand, if it consumes shrubs with mature leaves and with a high cellular wall content, the absorption of water will be less. Therefore, preserving and encouraging the establishment of alternative water sources, such as wild plants that provide succulent flowers and fruits consumed by white-tailed deer, is an important strategy to conserve and manage the deer in xerophilous thickets of northeastern Mexico. Also, in regions where the deer has access to plants of the *Opuntia* genus such as cactus, which contains around 90% humidity; however, cactus water cannot be completely absorbed because the cactus contains polysaccharides such as pectin, which tends to form gels in the digestive system. These gels retain water and reduce absorption in the intestine.

The small intestine of the deer absorbs from one to two liters day^{-1}. In addition, he must receive five to seven liters of water from secretions of the salivary glands, stomach, pancreas, liver and the same small intestine. While the intake enters the large intestine, approximately 80% of water has been absorbed. The net movement of water through cell membranes is usually carried out by osmosis. The absorption of water is dependent on the absorption of solutes, particularly Na. That the water flows from a low concentration of solutes to a solution high in solutes, by means of the following mechanisms dependent on the Na pump:

1. Na is absorbed into the cell by several devices; The most efficient of these is the transport that includes glucose and amino acids.
2. Absorbed Na is rapidly exported from the cell via the Na pump because a considerable amount of Na is pumped out of the cell, which establishes high osmolality in the small intracellular spaces between adjacent enterocytes.
3. Subsequently, the water diffuses in response to an osmotic gradient established by the Na within the intracellular space. Apparently, most of the water absorbed is transcellular, but some of it diffuses through the narrow junctions.
4. Finally, water as well as Na diffuse into the capillary vessels within the villi of the intestine.

Therefore, water is absorbed within the intracellular space by a low osmotic gradient. However, if the process is seen, the transport of water from the intestinal lumen into the blood is sometimes carried out against an osmotic gradient. This is important because it means that the intestine can absorb water into the blood even though the osmolality in the lumen is greater than the osmolality in the blood. Therefore, the proximal small intestine functions as a highly permeable mixed segment and the absorption of water is basically isotonic; which means that water is not absorbed until the intake has been diluted just above the osmolality of the blood. Likewise, the ileum and especially the colon can absorb water against an osmotic gradient of several hundred millimoles.

Body water and replacement

The greater amount of water found in deer tissues is present in intracellular fluids (40% or more of the total deer weight). Most intracellular water is found in muscle tissue. Extracellular water is found in interstitial fluids, which occupy the spaces between cells, blood plasma, and other fluids such as lymph, synovial fluid, and cerebrospinal fluid. Extracellular water occupies approximately one third of the total deer water. The rest of the body water is found in the digestive and urinary tracts. As mentioned above, the amount of water present in the digestive system varies greatly as it depends on the diet of the deer. Body water tends to decrease with the age of the deer and maintains an inverse relationship with the body fat content. That is, the fawns contain more water in their tissues than the adult deer; In addition, deer fasted by different circumstances that have made them lose weight (fat reserves) have more water than those that are not fasted because the water content in the body is inversely proportional to the body fat content.

Sources and losses of water

The deer can obtain water from five sources: 1) surface water or free water, such as that found in ponds, streams and dew that accumulates in plants, 2) water from food, 3) metabolic water, produced by the oxidation of organic nutrients, 4) preformed water that is associated with bodily tissues that are catabolized during periods when the energy balance of the deer is negative (long fasts, when the deer has to catabolize their fat reserves to produce energy), 5) water that is released from reactions where amino acids are condensed into peptides (mainly when the deer is in the growth stage). The amount of water that the deer receives from the last three sources, which are mentioned above, depends on their diet, habitat and ability to conserve body water.

The amount of water that the deer requires depends on several factors: 1) environmental temperature; 2) type of food; 3) physiological state and 4) the activity that is developing. However, apparently 3 to 6 liters of water per day are necessary to meet their demands. The amount of water consumed by the deer is inversely proportional to the concentration of water in the food. Although it has not been established experimentally, if the deer can survive without drinking water and rely exclusively on succulent and abundant green forages. The information available on the water needs for the deer is incipient. It has been suggested that in temperate climates the consumption of water by deer is 2 to 3 times the volume of dry matter they consume. Water consumption ad libitum (to free access) increased with the increase in the ratio of dry guajillo containing some secondary compounds, and dry ground alfalfa in diets of adult males. The daily water consumption was of 6.7, 6.2, 5.4, and 4.4 liters with 0, 25, 50, and 75% of guajillo, respectively. The minimum and maximum estimated water consumption in Jalisco, Mexico was 1.9 to 3.9 liters per adult deer, 1.4 to 1.7 liters per juvenile deer, and 0.8 to 1.7 liters per fawn. In contrast, female mule deer in Arizona require 5.0 to 5.5 liters of water in a 24-hour period. Deer reduce dry matter consumption when water consumption is restricted to 33% of their consumption at free access

Fodder plants usually contain abundant amounts of water of 45 to 65% in forage shrubs and 70 to 90% in herbs, fruits and flowers. The nopal (*Opuntia engelmannii*) is, especially an important source of water (90%) for the deer in northeastern Mexico and south Texas, USA. Drinking water is very important in these regions especially during hot summers when the temperature commonly reaches 38 to 43 ° C. Water availability can be quite critical during the dry season when herbs and other succulent vegetation is scarce.

The deer loses water through: 1) urine; 2) feces and 3) insensitively, which is water that is lost through vaporization in the lungs and dissipation through the skin and sweating through the sweat glands, when the temperature is warm. Losses through urine and feces occur at intervals, but losses through the lungs and skin are constantly present.

Water restriction

During hot summers or during drought when succulent forages are scarce, the deer does not receive sufficient quantities of water for maintenance, which can lead to water restriction. Deer predators such as the coyote and puma can scare the deer from water sources and suffer from restriction. The most noticeable effect in deer with moderate water restriction is the reduction of food consumption and productivity. Urinary and fecal excretions decrease considerably.

In the case of a more severe restriction, a rapid loss of weight is shown as the deer is dehydrated. Dehydration is accompanied by an increase in renal excretion of N and electrolytes, such as Na and K. Water restriction causes acute and rapid responses when temperatures are extremely high, as could happen in northeastern Mexico and southern Mexico. Texas, USA, when in the summer they can reach up to 45° C. In addition to dehydration, they usually increase:

1. the frequency of the pulse and rectal temperature, this happens when the deer no longer has enough water that can evaporate to maintain your normal body temperature
2. the respiratory rate
3. the concentration of water in the blood
4. nausea
5. difficulty in muscle movements and
6. death in case the water deficiency continues for a prolonged period of time.

Chapter 4

Habitat of white-tailed deer

Abstract.-The habitat of the white-tailed deer is based on the availability of food, vegetation, water and space available. Habitat quality is defined as the ability of the environment to provide appropriate conditions for the persistence of the deer and its population. The quality of the soil, which affects the nutritional quality of the food, also plays an important role. These conditions influence the size of the population and the carrying capacity of the herd. Nutritional ecology is the science that relates an animal to its environment through nutritional interactions. The quality of the plants that the deer consumes can be improved through the manipulation of the vegetation (management of shrub vegetation). However, the grassland manager has much less control over the quality of the forage than over the quantity. The two habitat factors that most affect the diet and nutrition of the deer are the availability of plants (quantity and accessibility) and quality (nutritional content and digestibility). To maintain a good condition of the deer and its population, it is important to maintain a diversity of forage species for the nutrition of white-tailed deer. Under the same carrying capacity of a range, competition for fodder between deer and beef cattle is less than competition between deer and goats or sheep. The strategy of supplementing deer is not financially profitable.

Introduction

The habitat is the place where an organism that interacts with other organism lives. The habitat of the white-tailed deer must provide the necessities for the reproduction and the maintenance of the population. Changes in environmental conditions modify the composition of the diet, the use of cover, space and water by white-tailed deer. The most important environmental factors affecting these parameters include climate, especially quantity and distribution of precipitation and temperature, as well as chemical and physical properties of soils. Deer have specific habitat needs, so their distribution and abundance are limited to an area given by the quantity, quality and heterogeneity of available resources. These characteristics have an important influence on the size of the family environment, which, in turn, is determined by factors such as: population density, quantity and quality of the diet, plant cover, presence of water sources and reproductive activities. So, it is expected that environmental and reproductive conditions modify the size of the family environment.

Habitat characteristics

The habitat of the white-tailed deer is based on the availability of food, vegetation, water and space available. Habitat quality is defined as the ability of the environment to provide appropriate conditions for the persistence of the deer and its population. Habitat conditions influence the size of the population or the carrying capacity of the herd. Soil fertilization, which affects the quality of the food, also plays an important role. The riparian soils on the banks of rivers and streams are the richest in nutrients; consequently, these soils support relatively high populations of deer because they consume forage rich in nutrients.

The plant cover

It is an important protective component in deer habitat because it is associated with seasonal changes. Adequate coverage provides protection from bad weather and from predators (including humans). It also provides a bed and shelter where the deer feels safe. Shrubby plants (shrubs) in riparian soils, high-altitude grasslands provide these conditions. However, the hills and cleared areas do not provide good protection for the deer.

The food

It is in many circumstances the most crucial element in the habitat of white-tailed deer. The white-tailed deer in good condition with an approximate weight of 70 kg consumes around 2.0 kg (3% of its live weight) of dry matter per day. Like all ruminants, the deer processes the food through its stomach (rumen, reticulum, omasum and abomasum) and in its approximately 22 m of intestines. The food takes about 24 to 36 hours to be digested and go through all the digestive treatment. Its diet consists mainly of leaves and shoots of many climbing vines, green and succulent herbs, grasses, acorns, fungi, aquatic plants and other types of plant parts that have a height of about 1.5 m. Due to its selective habits, the deer can change its selectivity for the species of plants present in the area. As their favorite foods become less available, their diets are gradually changed to those less nutritious and less preferred which can produce adverse effects on their reproduction.

The water

Water places in the ranch are an important component of the habitat of the white-tailed deer because they are sites of frequent visit, therefore, the presence or absence of water can necessarily affect their daily activity. However, the white-tailed deer can survive without water from for relatively long periods of time, as long as there are succulent plants such as cacti. Water requirements vary seasonally, being higher during the summer months and lower during the winter. Therefore, for the deer to develop in optimal conditions, the components of its habitat food, cover and water should be in the best possible conditions.

The space

The usable space is the portion of the habitat that is, or can be used by, the white-tailed deer. However, not all the space in the range is useful as a habitat for the deer. For example, grassland pastures without shrub vegetation, growing areas and areas without vegetation. The optimal habitat for white-tailed deer is that which has a high degree of juxtaposition of types of coverage distributed evenly throughout the range, including open areas, and thermal and protective cover. Juxtaposition refers to the proximity of food, cover, and water. In an ideal situation, the spatial distribution of the food, cover, and water in the habitat should be uniform, it is not desirable for the resources to be concentrated in a part of the habitat. The pastures with open areas dominated by herbaceous species and grasses with good distribution and surrounded by shrub vegetation provide the optimal habitat for white-tailed deer. The deer prefer open spaces without bushes or almost without them among shrubs as feeding areas. The open spaces between shrub vegetation are central to the nocturnal feeding habitat.

Availability of native forage plants

Nutritional ecology is the science that relates an animal to its environment through nutritional interactions. Energy and nutritional requirements, browse and digestive efficiencies, abundance and type of food provide cause-and-effect relationships that determine the animal's body condition, changes in body mass and, ultimately, reproduction and survival. Many of those relationships are physiologically and quantitatively predictable. Nutritional ecology, therefore, offers a perspective of a general, quantitative and predictable theory of the key relationships between an animal species and its habitat.

The availability of nutritionally adequate food, appropriate plant cover, convenient fresh and clean water, and enough space, establish the quality of a deer habitat. These conditions influence the size of the population and the carrying capacity of the herd. The quality of the soil, which affects the nutritional quality of the food, also plays an important role. Rapid changes and the availability of adequate habitat, caused by climate changes, which are common in northeastern Mexico and southern Texas, USA, can, at a given time, reduce populations of white-tailed deer.

The two habitat factors that most affect the diet and nutrition of the deer are the availability of plants (quantity and accessibility) and quality (nutritional content and digestibility). Seasonal changes cause plants to vary in abundance, growth status and nutritional characteristics. The deer will always try to maintain a quality diet that meets their nutritional needs, adjusting the components of the diet as forage plants change quality. If one or both factors mentioned above are limiting, they will cause a detrimental effect on the nutrition of the deer.

A good habitat for white-tailed deer that occurs in northeastern Mexico and southern Texas, USA (Figure 4.1.) Must contain four primary categories of plants that are available for consumption by deer: 1) shrubby, 2) herbs, 3) pastures and 4) cacti. However, the proportion of each category or of a species in the diet of the deer will vary, between years, seasons, regions and between deer. The availability of a preferred plant is a key factor that contributes to this variation. When the plant is green and growing, the deer will consume it (at least in small proportions). However, if a category of plants (herbs) is not very available in the area where the deer graze, then their diet will reflect a higher than normal preference, for the category of shrub plants, cacti and possibly native grasses. Similarly, if a species is not present in the deer's habitat, obviously this species will not appear as a preferred plant for the deer.

Figure 4.1. group of plants that consume the white-tailed deer in northeastern México and south Texas, USA.

In most of the farms of northeastern Mexico, the availability of forage is not usually a problem as shown in Table 4.1, which lists the species of plants that grow in the pastures of the counties of Anahuac, Parás, Vallecillo and Linares of the state of Nuevo León, Mexico, belonging to the Tamaulipan Thorny Scrub of the Gulf Coastal Plain and, for the most part, are consumed by white-tailed deer, where shrubs, grasses and grasses represent 68.0, 13.0 and 19.0%, respectively. The list of plants consumed by the deer is shown in Table 4.1.

Table 4.1. Botanical composition by groups of plants found the diet of white-tailed deer in northeastern Mexico

Trees and shrubs	Forbs	grasses
Acacia berlandieri Benth.	*Abutilon parvalum* Gray.	*Arsitida* spp
Acacia farnesiana L.	*Agrythamnia neomexicana*	*Buteloua gracilis* H.B.K.
Acacia rigidula Benth	*Aphanostephus* sp.	*Bothriochloa annulatum* Kuntze
Acacia wrightii Benth	*Arthemisa mexicana* L.,	*Cenchrus ciliaris* L.
Aloysia gratisima Gill and Hook	*Coldenia greggii*,	*Cenchrus incertus* M.A. Curtis.
Atriplex spp	*Cynanchum barbigerum* Scheele.,	*Chloris ciliata* Sw.
Bernardia myricaefolia ((Scheele)Wats.	*Dalea pogonatera* Gray.	*Digitaria californica* (Benth) Henr.
Bumelia celastrina HBK	*Dyssodia pentachata* (DC) Robins	*Hilaria berlangeri* Steud.
Caesalpinia mexicana A.Gray.	*Dyssodia micropoides* DC.,	*Panicum hallii* Vasey.
Calliandra conferta Gray	*Haplopappus spinolosus* (Greene) Hall	*Setaria macrostachya* H.B.K.
Cassia greggii Gray	*Heliotropium angispermum* Murr.	*Tridens muticus* (Torr) Nash.
Castela texana T. And G. Rose. (F)	*Hibiscus cardiophyllus*	
Celtis pallida Torr.	*Hibiscus* sp.	
Cercidium macrum I.M. Johnst	*Oxalis dichondrifolia* Gray.	
Condalia obovata IM Johnst	*Palafoxia texana* DC.	
Cordia boissieri A.DC.	*Polyanthes maculosa* Hook	
Desmanthus virgathus L.,	*Rhus* sp.	
Diospyros texana Scheele. (F)	*Ruellia corzoi* Tram & Burlk	
Ephedra aspera Engelm.	*Sida filicaulis* T. & G.	
Eysenhardtia polystachya (Ortega) Sarg.	*Solanum eleagnifolium*	
Forestiera angustifolia Torr.	*Verbena* sp.	
Gymnosperma glutinosum (Spreng) Less.	*Zexmenia hispida* H.B.K.	
Helieta parvifolia (Gray) Benth.	*Zephyranthes* spp	
Jatropha dioica Cerv.		
Karwinskia humboldtiana (R and S) Zucc.		
Krameria ramosissima (Gray.),		
Lantana macropoda Torr.		
Larrea tridentata DC.		
Leucaena leucocephala L.		
Leucophyllum texanum Berl.		
Lycium berlandieri M. Dunal		
Opuntia engelmannii Engelm.		
Opuntia leptocaulis A. P. de Candole		
Opuntia leptoculis DC.		
Parkinsonia aculeata L.		
Pithecellobium ebano (Berl) Muller.		
Pithecellobium pallens (Benth) Standl.		
Porlieria angustifolia Engelm.		
Prosopis glandulosa Torr.		
Prosopis glandulosa Torr. (F)		
Schaefferia cuneifolia Gray.		
Yuca sp		
Yuca sp. (F)		
Zanthoxylum fagara (L.) Sarg.		
Ziziphus obtusifolia T and G.		
83 % = composition of the annual diet	17 % = composition of the annual diet	1 % = composition of the annual diet

The deer select a greater number of bushes (83.0%), followed by native grasses (17.0%) and finally native grasses (1.0%). However, the exception is when the bushes have been eliminated and have been replaced by meadows with introduced or naturalized grasses such as *Cenchrus ciliaris* grass. However, communities with mixtures of shrubs and grasses provide the deer with sufficient quantities of moderate to high-quality nutritive shrubs.

Herbs, on the other hand, are generally scarce. High-quality perennial herbs are not very common due to improper pasture use. Annual herbs are highly dependent on soil moisture and are usually present only in brief periods during summer and autumn. Not very cold winters with adequate humidity result with a good emergence of temperate climate herbs.

The diversity of forage plants is an important component of the range for the nutrition of white-tailed deer. The diversity of edible and nutritious plants, allows the deer to select a quality diet, from these available species that as the seasons fluctuate, they also vary in their nutritional quality. Diversity is particularly important if the species manifest different states of growth during the seasons of the year. A large variety of plants in different growth stages increases the probability, throughout the year, of the availability of high nutritional quality.

Deer can change the components of their diet in response to changes in nutrient levels associated with the seasonal growth of each species. Diversity is also important, when the fruits of the legumes, which in the ecosystems of northeastern Mexico and south Texas, USA, are produced at the end of winter and early summer, since they represent an important source of energy for the deer that can consume them to recover the energy losses suffered during the mating. *Prosopis juliflora* pods are an important source of energy for white-tailed deer that grows in northeastern Mexico and southern Texas, USA.

Importance of diversity native plants

The quality of the plants that the deer consumes can be improved through the manipulation of the vegetation (management of shrub vegetation). However, the grassland manager has much less control over the quality of the forage than over the quantity. The quality of the forage is associated with the growth state of the plant, plant species and environmental factors such as soil type and precipitation. However, no plant can maintain, throughout the year, the level of nutrients required by the deer to obtain optimum growth and reproduction. Some species have higher nutrient content than other plants within the same group. An example is the *Celtis pallida* shrub that grows in northeastern Mexico, maintains adequate levels of crude protein, energy and minerals throughout the year. Some shrub species can maintain, throughout the year, adequate levels of a particular nutrient (such as crude protein) but may be seasonally deficient in energy or certain minerals required by the deer. The fodder cactus (*Opuntia engelmannii*) has a high digestibility of organic matter but is relatively low in crude protein, phosphorus and sodium. The foregoing emphasizes the importance of maintaining a diversity of forage species for the nutrition of white-tailed deer.

Competition for forage

Interspecific competition occurs when different species such as deer and domestic fauna compete for forage resources. Competition does not occur simply because two species are consuming the same type of forage plants. It is possible for sheep, goats, cattle and cervids to occupy the same range without competition, if the animals are present in low numbers and if there is diversity and abundance of native forage plants. When animals are present in small amounts, competition is usually minimal. Competition is severe only when the number of animals exceeds the forage supply, or the number of deer exceeds the carrying capacity of the habitat. Several years of overgrazing results in a decrease in the vigor of the plants, forage production and the potential of animal production. Over-utilization of the grazing land has a direct impact on deer habitat and on the nutritional quality of their diets. Under the same carrying capacity of a range, competition for fodder between cattle and beef cattle is less than competition between deer and goats or sheep. Because deer and goats are selective browsers their diets are very similar, but they vary when compared to other animal species. However, competition between deer and goats may be minimal, if the number of deer and goats is relatively low and forage availability is high.

Sheep tend to consume mainly grasses and forbs, therefore tend to compete with the deer for forbs, especially when they are scarce. Cattle can compete, to a small degree, with the deer for tender pastures, although they also consume some forbs and shrubs, especially those that lack thorns. In an over grazed pasture, cattle can compete with the deer for the remaining forbs and shrubs

Intraspecific competition (competition between individuals of the same species) is common in deer herds in some areas of northeastern Mexico and southern Texas where carrying capacity has been exceeded. This competition between deer can be significant in areas where predators have been eliminated, especially where there is little hunting pressure. Another factor that can contribute to overpopulation is the hunting policy that only allows hunting adult males (Figure 4.2). The result is usually a low male: female ratio, low hunting rate compared to the births of fawns, overpopulation and decrement in body condition.

Figure 4.2. Deer hunting

Methods to increase forage for deer

One way to try to increase the body condition and size of the antlers of the deer, is to artificially increase the load capacity of the range, through the complementation with commercial ingredients. However, feeding deer artificially is extremely expensive and counterproductive because supplementing the deer would result in an increase in the reproduction and survival of the fawns, which would lead to an increase in population, above the carrying capacity. In addition, with very few exceptions, the strategy of supplementing deer is not financially profitable.

A third method to minimize intraspecific competition is to manage the number of deer, allowing to hunt a number of deer that allows to have a population below the carrying capacity. This method requires continuous monitoring of the number of deer, body condition and seasonal habitat conditions. This management strategy is successful, if the availability and quality of the forage, within the habitat that makes up the herd of deer, are adequate. However, improvement of herd nutrition may not occur if the number of domestic livestock can take advantage of additional forage of high nutritional qualit

Recently, in some farms of northeastern Mexico and south of Texas, USA has opted for the elimination (clearing), in alternating strips, of certain areas of the scrubland. The presence of sunlight and moisture caused by precipitation, cause areas of relatively higher production of new growth plant material containing mainly herbaceous plants (herbs and grasses) and shrub sprouts, which are mostly highly edible and of high nutritional quality for the deer. In this way, the body weight of the deer and its antlers could be increased, and the hobby of deer hunting is used more efficiently.

Capítulo 5

Proteins

Abstract.- Virtually all deer cells synthesize proteins to partially maintain their vital functions; Without protein synthesis, life could not exist. Ruminants such as deer that have a microscopic intestinal flora can synthesize proteins from non-protein nitrogen sources. Proteins are formed by simple units such as amino acids and have a great diversity in their chemical composition, physical properties, size, shape, solubility and biological functions. The concentration of N in plants is influenced mainly by three factors; 1) the supply of available N of the soil; 2) differences between grasses and legumes and 3) the state of maturity of the plant. A part of the protein of the diet of the deer is hydrolyzed during the first stage of the digestion by the microbes of the rumen, another part remains intact until it reaches the small intestine. The protein in the diet that escapes ruminal fermentation and is transported to the lower digestive tract is called a bypass protein or escape protein, to differentiate it from the protein synthesized by ruminal microbes and endogenous secretions. The nutritional requirements of white-tailed deer protein are not known for sure or are not well documented as is the case of domestic ruminants. Some physiological processes that involve nutrition, such as filling capacity with respect to their body size, appear to be different in deer compared to domestic ruminants. It is likely that few plant species that can maintain, throughout the year, the level of crude protein required by the deer for proper growth and reproduction.

Introduction

Proteins are organic compounds formed by amino acids linked by peptide bonds, involved in various essential vital functions, such as metabolism, muscle contraction or immune response. The protein metabolism in the rumen of the deer is extremely complex; microorganisms degrade food, initially destroying the cell wall and initiating the continuous hydrolytic process of proteins. Protein destruction by fermentative deamination produces carbon dioxide, ammonia and short chain fatty acids. The amino acids, urea and nitrates, are converted into ammonia, used by microorganisms to synthesize amino acids and eventually convert them into proteins; a part of the ammonia is absorbed in the rumen, passes into the blood and is excreted in the urine in the form of urea and another is recycled by the saliva and ruminal walls.

Classification and functions

Proteins are organic substances or compounds essential for deer and is the class of nutrients that are found in a high concentration in deer muscle tissue. Virtually all deer cells synthesize proteins to partially maintain their vital functions; Without protein synthesis, life could not exist. Ruminants such as deer that have a microscopic intestinal flora can synthesize proteins from non-protein nitrogen sources. All deer cells contain proteins and cell turnover takes place very quickly in some tissues, such as the epithelial cells of their intestines. The protein content that is required in the diet of the deer is much higher in young animals in the growing stage and decreases as it reaches adulthood, when only a sufficient amount of protein is required to maintain body tissues. The different physiological states of the deer, such as pregnancy and lactation, increase the need for protein. Proteins are performed by simple units such as amino acids and have a great diversity in their chemical composition, physical properties, size, shape, solubility and biological functions (Figure 6.1). They perform functions as varied as cell membrane components to be part of the globulins that provide or give immunity to animals.

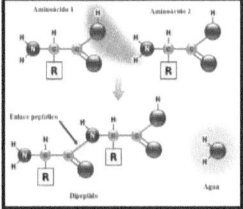

Figure 6.1. Basic structure of amino acids and peptidic bond to form proteins

Table 6.1. Amino acids and their functions found in body tissues of white-tailed deer

Amino acid	Function
Aspartic acid	It is very important for the detoxification of the liver and its correct functioning. L-aspartic acid combines with other amino acids forming molecules capable of absorbing toxins from the bloodstream.
Alanine	Involved in the metabolism of glucose. Glucose is a simple carbohydrate that the deer uses as an energy source.
Glutamic acid	It has great importance in the functioning of the central nervous system and acts as a stimulant of the immune system.
Asparagine	The nervous system requires asparagine. It also plays an important role in the synthesis of ammonia.
Cysteine	Along with L-cystine, cysteine is involved in detoxification, mainly as an antagonist of free radicals. It also helps maintain hair health due to its high sulfur content.
Glycine	In combination with many other amino acids, it is a component of numerous body tissues.
Glutamine	Brain nutrient and intervenes specifically in the use of glucose by the brain.
Proline	It is involved in the production of collagen and has great importance in the repair and maintenance of muscle and bones.
Serine	Along with some amino acids, it intervenes in the detoxification of the organism, muscle growth, and metabolism of fats and fatty acids.
Tyrosine	It is a direct neurotransmitter and can be very effective in the treatment of depression, in combination with other essential amino acids.
Arginine	It is involved in the conservation of the balance of nitrogen and carbon dioxide. It also has a great importance in the production of growth hormone, directly involved in the growth of tissues and muscles and in the maintenance and repair of the immune system.
Phenylalanine	Involved in the production of collagen, mainly in the structure of the skin and connective tissue, and also in the formation of various neurohormones.
Histidine	In combination with growth hormone and some associated amino acids, they contribute to the growth and repair of tissues with a role specifically related to the cardiovascular system
Isoleucine	Involved in the synthesis of hemoglobin and maintains the balance of blood glucose. Involved in the production of energy and repair of muscle tissue.
Leucine	Together with L-Isoleucine and growth hormone it intervenes with the formation and repair of muscle tissue.
Lysine	It is one of the most important amino acids because, in association with several other amino acids, it intervenes in various functions, including growth, tissue repair, immune system antibodies and hormone synthesis.
Methionine	It collaborates in the synthesis of proteins and constitutes the main limiting factor in the proteins of the diet. The limiting amino acid determines the percentage of food that will be used at the cellular level.
Threonine	Together with L-Methionine and Aspartic acid it helps the liver in its general detoxification functions.
Tryptophan	It is involved in the growth and hormonal production, especially in the function of adrenal secretion glands. It also intervenes in the synthesis of serotonin, a neurohormone involved in relaxation and sleep.
Valine	It stimulates the growth and repair of tissues, the maintenance of various systems and nitrogen balance.

The proteins of all living beings are determined mainly by their genetics (with the exception of some antimicrobial peptides of non-ribosomal synthesis), that is, the genetic information determines to a great extent what proteins a cell, a tissue and an organism have. Proteins are synthesized depending on how the genes that encode them are regulated. Therefore, they are susceptible to external signals or factors. The set of proteins expressed in a certain circumstance is called a proteome.

The primary structure of the proteins is determined by the amino acid sequence in the protein chain, that is, the number of amino acids present and the order in which they are bound, and the amino acid sequence is specified in the DNA by the nucleotide sequence. There is a conversion system, called a genetic code, that can be used to deduce the first from the second. The secondary structure of proteins is the folding that the polypeptide chain adopts thanks to the formation of hydrogen bonds between the atoms that form the pepitic bond. The hydrogen bridges are established between the stable ones.

- *Alpha helix*: In this structure the polypeptide chain spirals on itself due to the turns produced around the alpha carbon of each amino acid. This structure is maintained thanks to intrachain hydrogen bonds formed between the -NH group of a peptide bond and the group -C = O of the fourth amino acid that follows it.
- *Beta sheet*: When the main chain is stretched to the maximum allowed by its covalent bonds, a spatial configuration called beta structure is adopted.
- *Beta turns*: Sequences of the polypepetric chain with alpha or beta structure, are often connected to each other by means of so-called beta turns. They are short sequences, with a characteristic conformation that imposes a sharp 180 degree turn to the main chain of a polypeptide
- *Collagen helix*: It is a variety of secondary structure, characteristic of collagen, protein present in tendons and connective tissue; It is a particularly rigid structure.
- *Beta sheets* or *folded sheets*: some regions of proteins adopt a zigzag structure and associate with each other, establishing bonds through interchain hydrogen bonds. All the peptide bonds participate in these cross-links, thus conferring great stability to the structure. The beta form is a simple conformation formed by two or more parallel

polypeptide chains (which run in the same direction) or antiparallel (which run in opposite directions) and are attached tightly by means of hydrogen bonds and various arrangements between free radicals of the amino acids.

The tertiary structure of proteins is the way in which the polypeptide chain folds in space. It is the arrangement of domains in space. The tertiary structure is carried out in such a way that the apolar amino acids are placed inwards and the polar ones towards the outside. Likewise, the tertiary structure of proteins is stabilized by covalent bonds between cystine, hydrogen bonds between side chains, ionic interactions between side chains, van der Waals interactions between side chains and the hydrophobic effect (exclusion of water molecules, avoiding its contact with hydrophobic residues, which are packaged inside the structure).

The quaternary structure is the spatial arrangement of the different polypeptide chains of a multimeric protein, that is, composed of several peptides. It includes the range of oligomeric proteins, that is, those proteins that consist of more than one polypeptide chain, in which there may also be an alosterism behavior according to the concerted method of Jacques Monod.

The quaternary structure derives from the conjunction of several peptide chains (proteins) that, associated, make up an entity, a multimer, which has properties different from that of its component monomers. These subunits are associated with each other through non-covalent interactions, such as hydrogen bonds, hydrophobic interactions or salt bridges. In the case of a protein consisting of two monomers, a dimer, this can be a homodimer, if the constituent monomers are the same, or a heterodimer. There may also be disulfide bridge type bonds between cysteine residues located in different chains.

Deer proteins are classified according to their shape and solubility in water, salt, acids, bases and alcohol, and a classification as broad as this includes the following:

Globular. - They are soluble in water, diluted acids, bases and alcohol; for example: albumins, globulins, glutelins and prolamins.

Fibrous.- are insoluble in water, resistant to digestive enzymes; for example: collagen (animal protein more abundant in nature), elastins and keratins.

Conjugated.- are those that contain non-protein compounds, are of two types: lipoproteins such as deer cell membranes and glycoproteins such as cartilage, tendons and mucoproteins.

The functions of proteins in the deer organism are the following:

- Intervene in cell mobility.
- Many hormones are protein.
- Most enzymes are proteins.
- They are essential for the action performed by vitamins.
- They are part of the hormonal receptors.
- Some are second messengers for hormonal action.
- Form complexes with carbohydrates and lipids to form glycoproteins and lipoproteins.
- Participate in the immunological defense. Like immunoglobulins and complement system.
- Participate in muscle contraction.
- Proteins associated with buffer systems.
- Transport proteins such as albumin, hemoglobin and transferrin.
- Coagulation proteins.
- Regulatory proteins like cytokines
- Support proteins like collagen

Protein in plants that the deer consumes

In most plant tissues, nitrogen (N) is found in concentrations ranging from 1 to 5% (dry basis). These concentrations are high compared to other nutrients (except for K, which is the most abundant element in plants), possibly reflected by the need for N as a component of proteins, nucleic acids, chlorophyll and other minor constituents of the plants proteins. A large proportion of the protein in plants is of enzymatic origin, of which about half corresponds to the photosynthetic enzyme: ribulose diphosphate carboxylase. In general, the proteins of grasses and legumes are similar in what corresponds to their amino acid composition. The amino acids glutamic acid, aspartic acid and arginine are the main components of the proteins of grasses and legumes, being also the lysine, alanine and glycine present in substantial amounts.

The concentration of N in plants is influenced mainly by three factors; 1) the supply of available N of the soil; 2) differences between grasses and legumes and 3) the state of maturity of the plant. In pastures the concentration of N declines sharply as maturity increases, possibly due to the relative increase in the cell wall and the relative decrease in cytoplasm. In legumes the decrease is less marked than in pastures and shows more variation between species. This can be seen more marked in alfalfa and other legumes, showing a decrease of approximately 3.5 to 5.0%.

Within the context of the nutritional value of forage for deer, the concentration of total N is sometimes expressed as crude protein, which is the concentration of N multiplied by 6.25. This factor is derived from the average concentration of N contained in the plants, which corresponds to 16%. However, because the calculation is based on total N, the crude protein includes the protein itself and other nitrogen compounds (non-protein N, NNP). Usually, between 70 and 90% of the total N in grasses and legumes is present as a true protein and between 10 and 30% is present in other forms, which include amino acids, peptides, amines and sometimes nitrates.

Proteins associated with the cell wall

The proteins in the leaves are higher in quality than those stored in the seeds, although they can form variable amounts of non-protein nitrogen. Up to 90% of the true protein in the leaves of alfalfa can form a complex of photosynthetic enzymes. The proteins of the leaf can be divided into cytoplasmic and chloroplastic, nucleoproteins and proteins associated with the cell wall. The latter are presented in smaller quantities. The protein associated with the cell wall is sometimes called extensin, because its function may be a link between the fiber (covalently linked to the polysaccharides) and, it is much less soluble; In addition, it is recovered in the neutral detergent fiber (NDF), as well as some cytoplasmic and chloroplastic proteins denatured by heat. The N content in the NDF of food is markedly increased by heating, which promotes the denaturation of albumins. The nitrogen in the acid detergent fiber (FDA) is relatively indigestible and is poorly used by the rumen microorganisms. The protein that is insoluble in the

NDF but soluble in ADF presents high digestibility. The foliage of the native shrubs, herbs and regrowth of the native grasses that make up the diet of the white-tailed deer (*Odocoileus virginianus* texanus), during the wet season in northeastern Mexico and south Texas, USA, it generally contains low cell wall levels and, therefore, the amount of protein associated to cell wall is low.

Nitrogen metabolism in the rumen

At the intestinal level, the degradation of proteins is similar in deer and in non-ruminants. The proteins and peptides are degraded to oligopeptides by the action of pancreatic proteolytic enzymes (trypsin, chymotrypsin and carboxypeptidase), then the oligopeptides are degraded by the oligopeptidases of the apical membrane of the enterocytes releasing amino acids di and tripeptides that are finally absorbed. However, unlike non-ruminants, the protein that reaches the intestine of the ruminant is different from that ingested with the diet because ruminal microorganisms degrade more than half the proteins consumed. They do this by means of membrane proteases that unfold proteins into peptides and some free amino acids, which are absorbed by ruminal microbes.

There is a wide variety of nitrogen compounds available for the microorganisms present in the rumen of deer. These compounds include: proteins of various nature, which vary in solubility and content of amino acids, nucleoproteins containing a wide variety of purine and pyrimidic bases, various non-protein nitrogen compounds, peptides, amino acids, ammonia, amides, amines, volatile amines, salts of ammonium, nitrites and nitrates, as well as compounds such as urea (which also enters the rumen through the bloodstream and saliva)

The general procedures of microbial nitrogenous metabolism in the rumen of deer are the following:

- Initially the enzymes proteinases and peptidases hydrolyze proteins to peptides and free amino acids.
- Subsequently, the amino acids are used for the synthesis of proteins and other microbial cellular constituents, such as components of the cell wall and nucleic acids.

- Then the amino acids are also catabolized in AGV and other acids, CO2 and ammonia.
- The urea of the diet or recycled enters the rumen to be hydrolyzed to ammonia by the action of the urease enzyme.
- Compounds such as nitrates are reduced to ammonia.
- Ammonia is used in the synthesis of microbial cellular components such as proteins and others.

Digestion and metabolism of the PC in the deer

A part of the protein of the diet of the deer is hydrolyzed during the first stage of the digestion by the microbes of the rumen, another part remains intact until it reaches the small intestine. The hydrolysis of proteins due to the action of protease enzymes and peptidases in the rumen results in the formation of peptides and amino acids, which are subsequently deaminated and decarboxylated, with the respective release of ammonia. The ruminal protease bacteria are bound to the cells and, therefore, their proteolytic activity is directly related to the cellular biomass. Bacteria must physically contact the substrate so that proteolysis can occur. Protozoa differ from bacteria in that they engulf and digest food and microbial protein. The degradation of the protein in the rumen of the deer sometimes exceeds the microbial requirements of ammonia and although it is not very well clarified but varies greatly with their diet. However, when the supply of food nitrogen is less than the needs of the deer, the endogenous urea supplies it in part, and the amount of nitrogen that enters the intestine is greater than the amount of nitrogen ingested. Normally, more than half of the protein in the diet is hydrolyzed in the rumen, and most peptides and amino acids are deaminated. Most of the ammonia, along with free amino acids and alpha keto acids (acetate, isobutyrate, methylbutyrate and isovalerate) is converted into microbial protein in the rumen.

Ammonia that is not transformed into microbial protein, diffuses through the rumen wall and is transported in the bloodstream to the liver, where it is converted into urea. Part of the urea, then returns to the rumen via saliva, or via the bloodstream. However, most of the urea is removed from the blood by the kidneys, which excrete it in the urine. In the rumen amino acids are absorbed in small amounts since most free amino acids are deaminated to give rise to branched chain volatile fatty acids, CO2 and CH4.

The level of branched chain volatile fatty acids in ruminal fluid is an index of the degradation of amino acids in the rumen, since these normally derive from the fermentation of valine, leucine, isoleucine and proline, and are known as isobutyric, isovaleric acid, 2-methylbutyrate and valerate acids. Branched chain VFAs are used by bacteria as growth factors.

In the second stage of digestion, the residual protein of the diet and the microbial protein synthesized in the rumen, are hydrolyzed in the small intestine, releasing small peptides and amino acids, which are absorbed into the bloodstream. The amino acids and peptides are subsequently used for the synthesis of proteins in body tissues or to form milk. The undigested protein passes into the large intestine, where a small portion is digested, although most of it is excreted in the feces. The optimal pH in the rumen for protein proteolysis of the diet and deamination of amino acids is 6 to 7.

The availability of nitrogen for the synthesis of microbial protein in the rumen of the deer depends on the protein source and its susceptibility to microbial degradation, even when the deer consumes foods with low level of N or nitrogen compounds bound to the cell wall, the synthesis of microbial protein is low which can be pathological for the deer. P is required for the synthesis of microbial protein because P is incorporated into phospholipids.

Rumen protozoa play an important role in the degradation of deer diet protein because protozoa have proteolytic activity. Protozoa use bacteria as a source of N. Also, protozoa may be able to obtain energy from the fermentation of the protein due to the release of ammonia when deer consume grains.

The synthesis of microbial protein requires a source of energy such as adenosine triphosphate (ATP) that is generated by the catabolism of carbohydrates. The amount of energy required by microorganisms is used for maintenance and growth. It has been recommended that for optimum bacterial growth, the deer is required to consume around 19.3 g of N per kg of digestible organic matter. Herbs that in the humid season grow in northeastern Mexico and south Texas, USA, and that are avidly consumed by white-tailed deer, are a good source of digestible organic matter highly available for the synthesis of bacterial protein.

When the protein and energy balance is correct, protein production and, therefore, microbial growth are favored. The fermentation of glucose with the consequent production of volatile fatty acids (VFA) is increased to fill the strong energy demands linked to a rapid growth of microbes.

However, when there is an excess of carbohydrates in relation to the protein, there is a large amount of energy but insufficient nitrogen to sustain an adequate protein synthesis, therefore the microbial growth is not optimal. The energy becomes inefficient as it is used to maintain cells that are not replicating, instead of being used for the synthetic processes of growing cells. Also, when there is a large amount of protein in relation to carbohydrates, there is a lot of nitrogen to sustain growth, but it is limited due to insufficient energy input. This forces organic matter to use amino acids to fill its energy requirements, instead of using them for protein synthesis. The growth rate of organic matter is low and AGV production is moderate.

The main extracellular nitrogen compounds used in the synthesis of ruminal microbial protein and other cellular constituents are ammonia, amino acids and peptides. It has been observed that peptides are incorporated more efficiently than free amino acids to bacteria. Most free amino acids cannot cross the cell wall of bacteria, whereas peptides containing between 4 to 20 or more amino acids are used quickly, like ammonia. Most of the amino acids such as glutamic and aspartic acids, are rather metabolized to AGV than incorporated into the amino acids of the microbial protein. A significant number of species effectively utilize free exogenous amino acids as a source of nitrogen and carbon, however, the exogenous peptide carbon and peptide nitrogen are more efficiently converted to bacterial protein in the rumen than the carbon and nitrogen of free exogenous amino acids.

The peptides enter as such the bacterial cell and are then hydrolyzed to amino acids before their subsequent metabolism occurs. Numerous preformed amino acids are catabolized with ammonia production and this is used as the main source of nitrogen. Ammonia is essential for a significant number of ruminal bacterial species. Even when diets contain amino acids and peptides in sufficient amounts for all microbial protein synthesis and adequate amounts of carbohydrates as an energy source, a considerable amount of the amino acids is catabolized to ammonia, CO_2 and acids. Most rumen bacteria can synthesize major cellular constituents, including proteins, using ammonia nitrogen and exogenous carbon sources, such as certain AGV, organic acids, CO_2 and carbohydrates.

Bypass protein

A diet protein that escapes ruminal fermentation and is transported to the lower digestive tract is called a bypass protein or escape protein, to differentiate it from the protein synthesized by ruminal microbes and endogenous secretions. These terms can be confused. The protein of the diet that passes to the abomasum consists of two fractions: 1) The protein that evades the attack of microorganisms in the rumen and that through the esophageal groove goes to the abomasum without completely mixing with the ruminal content, which is called bypass protein; this is a characteristic of young ruminants, and 2) the protein that resists microbial attack in the rumen, as a result of the competition between ruminal digestion rates and the passage rate, which is called escape protein.

A variety of chemical and physical modifications have been made to increase the escape protein of dietary protein sources, among these are: 1) treatment with formaldehyde, 2) with tannins, 3) with heat and 4) with formulation with food sources, such as foliage of native trees and shrubs with relatively high content of condensed tannins, which are naturally low in ruminal availability, thus generating some types of less soluble protein and less subject to proteolysis; a variable proportion of these complexes thus formed, are divided by the acidic conditions of the abomasum.

The foliage of some trees and native shrubs that are an important component in the diet of the deer, contain secondary compounds such as condensed tannins, which bind with the protein of the diet forming a certain amount of protein that escapes the ruminal degradation and, in general, the indigestible link between the tannins and the protein of the diet, is broken by the acidic conditions of the abomasum of the deer, releasing the protein to be digested and transformed into peptides and amino acids that are eventually absorbed in the small intestine. Another important function of condensed tannins is the role they play to manipulate the balance of microorganisms in the rumen of deer. The ability of secondary compounds in plants such as condensed tannins to reduce protozoan populations has important implications for the availability of protein for deer.

Protein requirements of deer

The nutritional requirements of white-tailed deer protein are not known for sure or are not well documented as is the case of domestic ruminants. Some physiological processes that involve nutrition, such as filling capacity with respect to their body size, appear to be different in deer compared to domestic ruminants. Therefore, caution should be used when extrapolating the requirements of goats and / or sheep to deer, although in some circumstances these data may represent the best information available to estimate deer requirements. Deer requirements should be considered based on seasonal changes since these correspond to the physiological changes in the herd of the deer such as gestation, lactation, growth and development of antlers and available nutrients of the range.

The requirements are expressed in terms of metabolizable protein, which takes into account, for any type of food, the degradability of the protein (PD) in the rumen and the reach that is achieved during the synthesis of microbial protein, together with the digestibility, which is carried out in the second stage of the digestion of the protein of the diet, which is not digested in the rumen. To calculate the metabolizable protein (MP) requirements for maintenance, protein losses in feces, urine and skin should be taken into account as desquamations.

It is recommended that the requirement of MP for maintenance (MPm), in winter where the deer demand less MPm, is 3.10 g MP/kg of live weight (LW) raised to 0.75 power ($LW^{0.75}$) and during the summer, where the deer requires more MPm, it is 4.19 g MPm/kg $LW^{0.75}$. However, if the forage consumption increases in winter, the requirement increases to 4.64 g MPm/kg $LW^{0.75}$. Therefore, the daily CP requirement for white tail deer maintenance is 8.4% with a dry matter consumption (CDM) of 55 g/kg $LW^{0.75}$. This is the same level of consumption that the MPm satisfies for a diet of approximately 70% digestibility and an ME content of 2.4 Kcal/g in the dry matter (DM). In summer with a daily DM consumption of 50 g/kg $LW^{0.75}$, the MPm requirement is theoretically covered. For summer, the deer will require 9.1% CP during the summer for maintenance.

The requirements of MP for weight gain (MPg) in females and males, in different physiological states are shown in Table 6.1.

Table 6.1. MP requirements for weight gain (MPg) of white-tailed deer

Physiological state	DWG, g/day	CP, % in DWG	MP$_g$, g/g	MP$_g$, g/kg LW$^{0.75}$
New born (0-21 days)	382	20	0.29	19
Weaned off (> 20 wicks)	310	25	0.35	9
One year	130	25	0.31	2
Two years	46	23	0.36	1
Lactation	150	37	0.53	3
No nursing	208	27	0.38	3
Adult males	358	23	0.29	2

DWG = daily weight gain.
CP = crude protein.
MP$_g$ = metabolizable protein for weight gain.
LW = live weight

The metabolizable protein for pregnancy (MPgesta) for a female with a weight of 50 kg with a fetus of 5.8 kg requires 88 g MPgesta/day or 4.66 g/kg LW$^{0.75}$ or 7.8 g/kg LW$^{0.75}$, if taken in account the NRC maintenance requirement (2007). For lactation (MPl) 38 g/day or 8.6 g/kg$^{0.75}$/day are required.

The antlers of the adult male deer contain 45% of crude protein, and the development and size of their antlers is directly related to the level of protein consumption. Apparently, the deer must consume 15% of crude protein for an optimal growth of their antlers and they are estimated to be 0.26 g MP/kg LW$^{0.75}$, with a maximum rate of 0.54 g MP/kg LW$^{0.75}$. Apparently, the requirements for antler growth are less than maintenance requirements. Because male fawns have a larger size and growth, they need to consume more protein than females. The replacement and maintenance of muscle tissue of adults requires protein, but proportionally in smaller amounts than required by the fawns. The protein required by young people slightly exceeds that of adults. During the last third of gestation and during lactation, adult females require proportionally more protein than adult

Seasonal availability of protein in the forage

The nutritive value and palatability of a forage depend on the season of the year and determine how important that forage is for the deer and how much protein it will provide; for example:

Those forages that are more nutritious and palatable in the *spring* such as those shown in Tables 6.2, 6.5 and 6.6 are important for:

- Help the deer recover from winter stress
- Provide the pregnant female with the nutrients required to support the development of the fetus during the last third and to initiate milk production
- Provide the adult male with the nutrients required for the start of antler development

Those forages that are more nutritious and palatable in *summer*, such as those shown in Tables 6.3, 6.6 and 6.7, are important for:

- Provide lactating females with the nutrients required to produce milk
- Provide nutrients for the growth of the fawns
- Support the future development of the antlers of the males

Those forages that are more nutritious and palatable in *autumn/winter* as those shown in Tables 6.4, 6.7 and 6.8 are important for:

- Provide nutrients for newborn fawns and their development
- Provide energy required to prepare for winter
- Replenish the energy lost during the mating period.

However, when the protein is limiting for animals, symptoms of deficiency can occur such as: anorexia, decreased growth, negative nitrogen balance, decreased feed utilization, reduced serum protein concentration, anemia, accumulation of fat in the liver, edema in severe cases, low weight of offspring at birth, decline in milk production and reduction in the synthesis of some enzymes and hormones. Death due to malnutrition can occur, if the deer consumes a diet with a content lower than 7% of crude protein for a relatively long period of time because the microorganisms in the rumen would suspend their growth, causing that the intake of food to be reduced.

It is likely that few plant species that in particular can maintain, throughout the year, the level of crude protein required by the deer for proper growth and reproduction. However, there are plant species that stand out for their high protein content and if they are available in the range they will provide the deer with good physical condition. Tables 6.3 and 6.4 show the seasonal variation in the content of crude protein (CP) and degradable protein (DCP) in the foliage of native shrub species that grow in northeastern Mexico and southern Texas, USA, and that are consumed by white-tailed deer. Even when the guajillo leaves (*Acacia berlandieri*) contain enough crude protein, during all seasons of the year, to satisfy the deer's needs for growth, reproduction and lactation; however, the availability of the protein (CP-DCP), to be metabolized by the microorganisms in the rumen, is very low; especially during seasons where precipitation is low. Apparently, only in summer the guajillo contains crude protein in sufficient quantities (10%) to satisfy the maintenance demands of the microbial flora in the deer rumen. Also, it is very likely that the deer consuming only guajillo would not survive for a long time due to the low availability of N for the microbes in the rumen of deer. The same effect could occur if the shrub *Acacia rigidula* was the only source of food. It is posible that the low ruminal availability of the protein in *A. berlandieri* and *A. rigidula*, is due to its high content of condensed tannins because tannins tend to form very strong complexes with dietary proteins and, it is the most important aspect to provoke antinutritional and toxicological effects, decreasing the palatability and digestibility of feed.

However, species of shrub plants, which grow in northeastern Mexico and southern Texas, USA, are consumed by the white-tailed deer such as: *Acacia farnesiana, Acacia wrightii, Celtis pallida, Cercidium macrum, Desmanthus virgathus, Eysenhardtia polystachya, Karwinskia humboldtiana, Leucaena leucocephala, Pithecellobium ebony, Pithecellobium pallens, Prosopis glandulosa, Zanthoxylum fagara* (Table 6.4) contain crude protein (equal and in some cases greater than 20%), during the four seasons of the year, to carry out satisfactorily physiological functions such as reproduction, gestation and lactation, and with a percentage of soluble crude protein (digestible) above the minimum requirements (7%) for the growth of the microbes that symbiotically inhabit their rumen (Table 6.5). Therefore, trees and shrubs represent an important forage resource for white-tailed deer, not only because of their high content and availability of crude protein, but because they are perennial and are consistently available and can be consumed by the deer for a long time of the year.

Table 6.3. Seasonal content of crude protein (% of dry matter) in native trees and shrubs that grow in northeastern Mexico and southern Texas, USA and that are consumed by Texas white-tailed deer

Scientific name	winter	spring	summer	fall	Average
Acacia berlandieri (L.) Wild.	19	20	26	21	22
Acacia farnesiana (L.) Wild.	22	21	22	20	21
Acacia rigidula Benth.	17	16	17	13	17
Acacia wrightii Benth.	23	24	21	19	22
Bernardia myricaefolia (Scheele) Wats.	17	17	17	17	17
Bumelia celastrina H.B.K.	18	18	15	17	17
Caesalpinia mexicana A. Gray.	13	14	18	14	14
Castela texana T. and G. Rose.	14	15	16	14	15
Celtis pallida Torr.	24	19	22	26	23
Cercidium macrum I.M. Johnst.	24	23	26	26	25
Condalia obovata Hook.	16	19	13	15	16
Cordia boissieri A. DC.	19	14	12	14	15
Desmanthus virgathus L.	22	18	19	22	20
Diospyros texana Scheele.	12	12	14	14	13
Eysenhardtia polystachya (Ortega)	21	23	17	22	21
Forestiera angustifolia Torr.	17	21	19	12	17
Gymnosperma glutinosum (Spreng) Less.	19	14	12	14	15
Helieta parvifolia (Gray) Benth.	13	12	14	12	13
Karwinskia humboldtiana (R and S)	21	26	18	18	21
Lantana macropoda Torr	15	16	19	16	17
Larrea tridentata DC.	17	18	17	18	17
Leucaena leucocephala L.	20	25	26	23	24
Leucophyllum texanum Berl.	10	14	13	12	12
Opuntia engelmannii Engelm.	5	4	6	5	5
Parkinsonia aculeata L.	17	19	22	17	19
Pithecellobium ebano (Berl) Muller.	22	15	25	23	21
Pithecellobium pallens (Benth) Standl.	22	18	25	19	21
Porlieria angustifolia Engelm.	20	18	17	15	17
Prosopis glandulosa Torr.	19	19	22	18	20
Schaefferia cuneifolia Gray.	12	15	18	14	15
Zanthoxylum fagara (L.) Sarg.	20	26	21	19	22
Ziziphus obtusifolia T and G.	19	16	14	16	16
Average	18	18	18	17	18

Table 6.4. Seasonal content of degradable crude protein (% of dry matter) in native trees and shrubs that grow in northeastern Mexico and southern Texas, USA and that are consumed by white-tailed deer

Scientific name	winter	spring	summer	fall	Average
Acacia berlandieri (L.) Wild.	13	13	16	13	14
Acacia farnesiana (L.) Wild.	11	10	9	8	9
Acacia rigidula Benth.	12	11	13	10	13
Acacia wrightii Benth.	9	10	7	6	8
Bernardia myricaefolia (Scheele) Wats.	4	5	6	6	5
Bumelia celastrina H.B.K.	6	7	6	8	7
Caesalpinia mexicana A. Gray.	8	9	12	10	9
Castela texana T. and G. Rose.	3	3	5	5	4
Celtis pallida Torr.	4	4	6	5	5
Cercidium macrum I.M. Johnst.	10	9	10	9	10
Condalia obovata Hook.	5	7	6	4	6
Cordia boissieri A. DC.	9	6	5	6	6
Desmanthus virgathus L.	10	8	7	9	9
Diospyros texana Scheele.	3	4	4	4	4
Eysenhardtia polystachya (Ortega)	6	5	7	7	6
Forestiera angustifolia Torr.	4	5	4	5	5
Gymnosperma glutinosum (Spreng) Less.	5	2	2	2	3
Helieta parvifolia (Gray) Benth.	2	2	2	3	2
Karwinskia humboldtiana (R and S) Zucc.	4	7	6	6	6
Lantana macropoda Torr	4	4	5	5	5
Larrea tridentata DC.	4	7	7	7	6
Leucaena leucocephala L.	8	10	9	8	9
Leucophyllum texanum Berl.	4	6	5	5	5
Opuntia engelmannii Engelm.	2	1	2	2	2
Parkinsonia aculeata L.	5	5	7	7	6
Pithecellobium ebano (Berl) Muller.	12	7	16	13	12
Pithecellobium pallens (Benth) Standl.	10	5	6	8	7
Porlieria angustifolia Engelm.	8	7	6	5	6
Prosopis glandulosa Torr.	7	6	9	6	7
Schaefferia cuneifolia Gray.	2	3	3	3	3
Zanthoxylum fagara (L.) Sarg.	3	9	6	6	6
Ziziphus obtusifolia T and G.	5	6	5	6	6
Average	7	6	7	7	7

Forbs, on the other hand, have annual growth so they are scarce in the range during most of the year because they are highly dependent on soil moisture. They are present in these semiarid regions (northeast of Mexico and south of Texas, USA) only for brief periods in summer and autumn. In addition, in wet winters and that are not very cold there may be availability of forbs for the consumption of deer. Apparently, due to its high nutritive value and digestibility, forbs are the group of plants most preferred by the white-tailed deer that lives in northeastern Mexico; However, due to its low availability in the pasture, during the dry seasons (winter and spring), the deer have to make greater use of the shrub plants to satisfy their nutritional requirements (Table 6.5).

Table 6.5. Seasonal requirements for crude protein of white-tailed deer (% of dry matter consumed) in different physiological states and average crude protein content (% dry matter) of grasses, shrubs and grasses consumed by the deer in northeastern Mexico and south Texas, USA

Physiological state	CP requirements			
	Winter	Spring	Summer	Fall
Fauns			16	16
Maintennce of adults	7 a 10	7 a 10	7 a 10	7 a 10
Fenales on gestation	17	17	17	
Females lacting			17	
Antler development			15	15
Mating	10	10	10	10
	CP content in plants			
Forbs	17	14	17	15
Native Trees and shrubs	18	18	18	17
Native and introduced grasses	7	10	8	10
Average	14	14	15	14

Table 6.6 shows the seasonal content of CP and DCP in 13 species of forbs that grow in northeastern Mexico and southern Texas, USA. They stand out for their high content of crude protein *Zephyranthes* spp., *Heliotropium angiospermum, Cynanchum barbigerum, Ruellia corzoi, Oxalis saidandrefolia, Sida filicaulis* and *Dalea pogonatera*. In addition, in summer and autumn, with the exception of *Dyssodia pentachyata* and *Polyanthes maculosa*, all the herbs found had sufficient levels of crude protein to satisfy

the needs of fawns and females in gestation and lactation (Table 6.5). Herbs had average values of 17, 14, 17 and 15% in winter, spring, summer and autumn, respectively. In general, all the herbs that appear in it had levels of crude degradable protein to cover, with a wide margin, the needs of the rumen microorganisms of the white-tailed deer, which lives in these regions.

Probably the presence and greater abundance of forbs in the summer and autumn seasons, plays a very important role for the survival and development of the white-tailed deer, which lives in these regions because its presence and abundance coincide with the birth of fawns, lactation and the start and development of the antlers of the male deer (Table 6.5). Native and introduced grasses represent the least important group of plants in the white-tailed deer diet in northeastern Mexico and southern Texas, USA. Although the deer consumes them in relatively low amounts in summer and autumn. The ecological importance of introduced and native grasses that develop in arid and semi-arid regions is due to their ability to withstand long periods of drought. But, as all plants are affected by weather changes; although there is no factor that impacts the quality of the forage as the maturity of the plant.

Table 6.6. Seasonal variation of crude protein content (CP, % of dry matter) and degradable CP (DCP, % of dry matter) in native herbs of the flora of northeastern Mexico and south Texas, USA and that are consumed by white-tailed deer

Forbs	Type of protein	Winter	Spring	Summer	Fall
Coldenia greggii (T and G) Grav.	CP	17	12	14	12
	DCP	9	7	8	5
Cynanchum barbigerum Scheele.	CP		19	16	16
	DCP				13
Dalea pogonatera Gray.	CP				16
	DCP				12
Dyssodia pentachyata (DC) Robins	CP	12	9	11	10
	DCP	8	5	7	7
Happlopapus spinolosus (Greene) Hall.	CP	15			15
	DCP	10			12
Heliotropium angiospermun Murr.	CP	20	21		19
	DCP	16	9		14
Oxalis dichoandraefolia Gray.	CP				17
	DCP				13
Palafoxia texana DC.	CP	12	15	19	11
	DCP	9		16	8
Polyanthes maculosa (Hook) Shinners.	CP	12	9	15	12
	DCP	11	7	14	11
Ruellia corzoi Tram and Burlk.	CP		16	19	15
	DCP		10	14	8
Sida filicaulis T and G.	CP			18	15
	DCP			15	12
Zephyranthes spp.	CP	30		23	21
	DCP	27		20	19

In the farms of northeastern Mexico and southeastern Texas, USA, in which a diversified livestock is developed where there are domestic livestock and wildlife, especially white-tailed deer, a common practice is to carry out selective clearing of trees and shrubs, either in alternating strips or in ecologically strategic areas. The clearing can propitiate the entrance of solar light and, when there is abundant humidity, especially during the seasons of summer and autumn, makes possible the growth of annual native grasses. This practice promotes the sustainable development, under extensive conditions, of domestic livestock and white-tailed deer because young pastures supplement their diets.

In Table 6.7. it shows the seasonal content of CP and DCP of 9 native grasses of annual growth and the introduced grass *Cenchrus ciliaris* consumed by the white-tailed deer that inhabits northeastern Mexico and south Texas, USA. Except for *Panicum hallii* and *Cenchrus incertus*, the other pastures have very low levels of crude protein, during most seasons of the year, for the requirements of white-tailed deer (Table 6.5). Apparently, the pastures are more abundant during summer and autumn, because it is when there is greater precipitation, therefore it is in spring and autumn they contain more crude protein (an average of 10%) and degradable crude protein.

Table 6.7. Seasonal variation of the content of crude protein (CP,%), degradable CP (DCP, %) in native pastures of the flora of northeastern Mexico and south Texas, USA and that are consumed by white-tailed deer

Grasses	Type of protein	Winter	Spring	Summer	Fall
Aristida spp.	CP	6	6	6	7
	DCP	2	3	3	4
Buteloua gracilis H.B.K.	CP	9	7	11	9
	DCP	5	4	6	5
Cenchrus ciliaris L.	CP	6	10	7	10
	DCP	3	5	3	7
Cenchrus incertus M.A. Curtis.	CP	7	13	6	9
	DCP	4	8	4	7
Chloris ciliata Sw.	CP		7		9
	DCP		5		6
Digitaria californica (Benth) Henr.	CP		12		13
	DCP		7		8
Hilaria berlangeri Steud.	CP	4	7	8	9
	DCP	2	4	5	6
Panicum hallii Vasey.	CP	9	14	13	12
	DCP	5	8	8	9
Setaria macrostachya H.B.K.	CP	11	11	8	11
	DCP	5	7	5	8
Tridens muticus (Torr) Nash.	CP				8
	DCP				5

However, only grasses such as *Panicum hallii, Digitaria californica, Cenchrus ciliaris, Cenchrus incertus* and *Setaria macrostachya* had levels of degradable protein in the rumen of the deer to promote microbial growth. Perhaps because in general native grasses contain low crude protein and higher fiber content, the deer prefers to consume shrubs and herbs that contain considerably more protein and less fiber. However, it is likely that the deer include pastures in their diets due to their high fiber content since the fiber is used to carry out normal function in their rumen because it stimulates rumination and insalivation, another of its functions is its buffer capacity in the rumen. Fiber is also involved in regulating the voluntary intake of deer feed.

Chapter 6

Carbohydrates

Abstract.- Carbohydrates are the main reservoir of photosynthetic energy of plants. The nutritional characteristics of carbohydrates, for the feeding of the deer, are variable and, depend on their monosaccharide components and their unions. Carbohydrates make up most of the cell wall of plants and as such, play an important role in the structural integrity of individual cells, tissues and organs. In deer almost, all carbohydrate digestion occurs within the rumen (more than 90%), although under certain circumstances, such as high rates of ruminal passage, a significant amount of carbohydrate digestion can occur in the small intestine and large intestine. The walls of plant cells can be considered as a compound, consisting of cellulose fibrils embedded within a matrix of lignin and hemicellulosic polysaccharides. In addition, the intact cell wall contains components such as water, organic and phenolic solvents, which give the structure unique properties. The most obvious function of the cell wall is its role in morphogenesis. The cell walls form the structural design of the plant architecture and provide mechanical and structural support for the plant organs. In addition, the walls play important roles in water balance, ion exchange, cell recognition and biotic stress protection. The cellulolytic bacteria that are found in the rumen of the deer, of which *Ruminococcus flavefaciens, R. albus* and *Fibrobacter succinogences* are the most important and are responsible for the cellulose digestion that occurs in the rumen. Ciliated protozoa and fungi that have also been identified in ruminal microbial populations have cellulolytic activity; however, its contribution to the degradation of cellulose is relatively minor. The degradation of hemicellulose in the rumen of the deer occurs in a manner analogous to that of cellulose but involves a broader arrangement of enzymatic activities. Some ruminal fungi and protozoa also have hemicellulolytic activity, but their activity in the degradation of hemicellulose is relatively minor, compared with bacteria.

Introduction

They are molecules composed mostly of carbon, hydrogen and oxygen atoms. They are very important components of the diet of the deer, so they represent the primary biological form of storage or energy consumption; in addition, of fats and proteins. In deer they are part of biomolecules isolated or associated to others such as proteins and lipids. They are not molecules whose carbons are hydrated but bound to alcoholic or hydroxyl groups (-OH), and to hydrogen radicals (-H). In addition, there is always a functional group such as a ketone group (-C = O-) or an aldheido group (-CH = O), so the carbohydrates could be called polyhydroxyketones (ketoses) or polyhydroxyaldheidos (aldoses). Types of important carbohydrates are: monosaccharides that cannot be hydrolyzed. Disaccharides upon hydrolyzing produce two monosaccharides. Oligosaccharides to hydrolyze produce three to ten molecules of monosaccharides. Hydrolyzing polysaccharides produce more than ten monosaccharide molecules. In the plants they perform diverse functions, being the one of energetic reserve and formation of structures the two most important ones. In addition, in the organism of the deer maintain the muscular activity, body temperature, blood pressure, the correct functioning of the intestine and neuronal activity. They represent the main molecules stored as a reserve in living beings along with lipids. They are the main substances product of photosynthesis and are stored in the form of starch in high amounts in plants.

Carbohydrates as components of the forage

Carbohydrates are the main reservoir of photosynthetic energy of plants. The nutritional characteristics of carbohydrates, for the feeding of the deer, are variable and, depend on their monosaccharide components and their unions. However, the variety of molecules and bonds in plants is much wider than that of animal tissues. The carbohydrates in plants contain many sugars and bonds not common to animal systems. Nutritional availability depends on the ability to break glycosidic bonds in plants and between carbohydrates and other substances. They form the bulk of the food supply for deer, and it is the most abundant class of components found in plants. They play important roles in the intermediary metabolism, energy transfer, storage and structure of the plant. Photosynthetic energy is fixed in carbohydrates via the Calvin cycle; In addition, they serve as initial substrates for almost all intermediate plant routes. Energy is transported within plants as the disaccharide sucrose and stored in polymers such as starch and fructans.

Anabolism of carbohydrates

In the anabolism of carbohydrates, simple organic acids can be synthesized from monosaccharides such as glucose and then synthesize polysaccharides such as starch. The generation of glucose from compounds such as pyruvate, lactic acid, glycerol and amino acids is called gluconeogenesis. Although fat is a common form of energy storage, in vertebrates such as deer, fatty acids cannot be converted into glucose by gluconeogenesis, since these organisms cannot convert acetyl-CoA to pyruvate. As a result, after a time of starvation, vertebrates need to produce ketone bodies from fatty acids to replace glucose in tissues such as the brain, which cannot metabolize fatty acids.

Carbohydrates are produced through the photosynthetic process of carbon fixation. The formation of different types of sugars, usually occurs through the action of the enzymes epimerases, isomerases, oxidoreductases and / or decarboxylases on activated monosaccharides that leave the Calvin cycle or the breakdown of storage carbohydrates. The biosynthesis of oligosaccharides and polysaccharides requires activated sugars, in the form of monosaccharides diphosphate nucleosides. The most predominant pattern is glucose interconversions, derived directly from photosynthetic activity or from starch degradation. There are a few alternative routes such as, the conversion of inositol or glucuronic acid and degradation pathways that can claim galacturonic acid and galactose through direct phosphorylation.

Among the most important carbohydrates in deer nutrition are the following:

Glucose.- It is the main product of starch digestion in norumiantes, it occurs in blood, lymph, cerebrospinal fluid and energy metabolism.

Fructose.- It is free in green plants, semen and blood of fetuses.

Manosa.- It does not exist free in nature, but it does form the mornings of yeasts, molds and bacteria.

Galactose.- It does not exist free and is a product of fermentations.

Sucrose.- Formed by fructose and glucose Also called sugar cane or beet, is for domestic use.

Lactose.- Composed of glucose and galactose, called milk sugar and is a product of the mammary gland.

Maltose.- Formed by 2 glucose linked by the bond α-1-4. Malt sugar is produced by hydrolysis of starch and malt.

Cellobiose.- It is a disaccharide present in cellulose composed of two molecules of glucose so it can be hydrolyzed to glucose.

Raffinose.- Formed by a glucose, a fructose and a galactose, is found in plants such as sugar beet and starch.

Starch.- Main nutritive reservoir in the plants that the deer consumes, composed of fractions of amylose and amylopectin.

Glycogen.- Highly branched polysaccharide of animal or microbial origin, found in the liver muscle and other tissues of the deer. When glucose cannot be stored as glycogen or converted immediately to energy, it is converted to fat. Glycogen is a polymer of α-D-Glucose identical to amylopectin, but the branches are shorter (approximately 13 units of glucose) and more frequent. The glucose chains are organized globularly like the branches of a tree originating from a pair of glycogenin molecules, a protein with a molecular weight of 38,000 that serves as a primer in the center of the structure. Glycogen is easily converted to glucose to provide energy.

Dextrins.- Intermediate products of glycolysis of starch and glycogen.

Cellulose.- Main constituent of the cell wall of plants, its function is more structural than nutritional. The final product of cellulose digestion is a mixture of volatile fatty acids and CH_4 and CO_2.

Fructosanas.- Fructose polymers that serve as a reserve for the plants that the deer consumes.

Hemicellulose.- Branched and linear polysaccharides with traces of sugars such as: xylose, arabinose, glucose, galactose and uronic acids. It is found in the cell wall of forage plants.

Pectin.- It is found in intercellular spaces formed by polysaccharide chains, substituted in galactana, arabana and galacturonic acid side chains.

Lignin.- It is not a carbohydrate, it is found in the cell wall, it is very resistant to acids and the action of microorganisms, it cannot be digested by animals, it represents 5% to 10% of the dry matter of plants. The lignin molecule has a high molecular weight, which results from the union of several phenylpropyl alcohols and alcohols (cumarilic, coniferyl and sinapyl). The random coupling of these radicals gives rise to a three-dimensional structure, amorphous polymer, characteristic of lignin. Lignin is the most complex natural polymer in relation to its structure and heterogeneity. For this reason it is not possible to describe a defined structure of lignin; however, numerous models have been proposed that represent its structure.

Carbohydrates make up most of the cell wall of plants and as such play an important role in the structural integrity of individual cells, tissues and organs. In deer almost, all carbohydrate digestion occurs within the rumen (more than 90%), although under certain circumstances, such as high ruminal passage rates, a significant amount of carbohydrate digestion can occur in the small intestine. and large intestine. The sugars are rapidly fermented in the rumen, fermentation produces VFA which are absorbed into the blood through the ruminal wall. The polysaccharides must be degraded into simple sugars before they are used. Non-structural polysaccharides such as starch and fructanas are rapidly and completely degraded within the rumen, while the degradability of structural polysaccharides (cellulose and hemicellulose) varies considerably and they are degraded more slowly and incompletely. The degradability of cellulose in forages varies from 25 to 90%, while the hemicellulose digestibility varies from 45 to 90%. The degradation of β-glucans is intermediate to cellulose. This ability to degrade and use structural carbohydrates gives ruminants a unique ecological niche. In addition to being the main source of energy in the diet of the deer, they have other important nutritional roles. Structural carbohydrates are important for normal rumen function; the fiber stimulates rumination, salivation and the property it has for the exchange of cations that are essential in the control and buffering of ruminal pH and fiber play and important role in voluntary intake of deer.

Catabolism of carbohydrates

In deer, the route of preferential metabolization of glucose involves the division of the molecule into two lactates. This metabolization or fermentation also takes place between many species of microorganisms and is characteristic of muscle cells. The muscle is a tissue in which fermentation represents a very important metabolic path since muscle cells can live for long periods of time in environments with low oxygen concentration. When these cells are actively working, their energy requirement exceeds their ability to continue with the oxidative metabolism of carbohydrates because the speed of this oxidation is limited by the speed at which oxygen can be renewed in the blood. The muscle, unlike other tissues, produces large amounts of lactate that is poured into the blood and returns to the liver to be transformed into carbohydrates. In the intermediary metabolism (Figure 6.1) are those of carbohydrates and the main ones are:

1. Glycolysis
2. Glycogenesis
3. Pentose cycle

Oxidative metabolism gives common routes with lipids such as the Krebs cycle and the respiratory chain. The main hormone that controls the metabolism of carbohydrates is insulin. In ruminants almost all carbohydrate digestion occurs within the rumen (more than 90%), although in white-tailed deer, which have high passage rates, a significant amount of carbohydrate digestion can occur in the small intestine and large intestine. The sugars are rapidly fermented in the rumen to give VFA which are absorbed into the blood through the rumen wall. The polysaccharides must be degraded into simple sugars before they are used. Non-structural polysaccharides such as starch and fructans are rapidly and entirely degraded, within the rumen, while the degradability of structural polysaccharides varies considerably.

The ability to degrade and use structural carbohydrates gives ruminants a unique ecological niche. In addition to be an important source of energy in the diet of ruminants, carbohydrates have other nutritional roles as components of dietary fiber. Structural carbohydrates are important for normal rumen function. The fiber stimulates rumination and salivation and, promotes the exchange of cations that are important in the capacity of ruminal buffering. Fiber is also involved in the regulation of voluntary consumption. In plants, cellulose is one of the carbohydrates that is part of its structure; in all forages its presence is high. In herbivorous animals, the bacterial flora is responsible for degrading it to obtain assimilable nutrients. In animals, the main energy reserve is glycogen, which is found in the liver and muscles. From the digestion of carbohydrates in ruminants the main products obtained are volatile fatty.

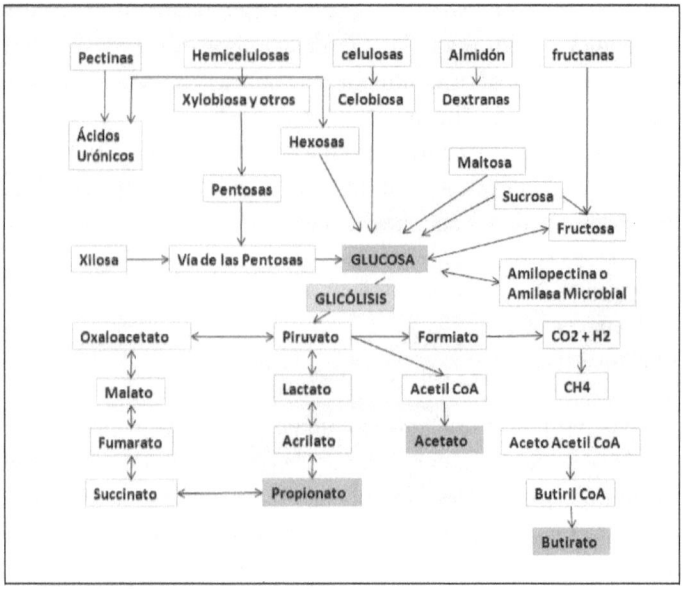

Figure 6.1. Metabolic pathways of carbohydrates

The Pentose Phosphate Cycle

This cycle aims to generate NADPH necessary for the synthesis of fat and generate ribosomes, takes place in the adipose tissue mammary gland, liver, adrenal cortex and thyroid. The pentose phosphate cycle is also known as the pentose deviation cycle, the hexose mono phosphate pathway or the oxidative pathway of phosphoglycerate.

Digestion of carbohydrates in ruminants

In ruminants there is a microflora and a micro fauna constituted by bacteria, fungi, protozoa, yeasts and viruses, which are contained in the ruminal reticulum. The population of bacteria is 1×10^{10} and that of the protozoa is 1×10^6 per milliliter of ruminal fluid. The main function of fungi is the breaking of the fiber so that the bacteria are more efficient. After the microflora and the micro fauna act, the presence of monosaccharides is abundant.

Glucolysis

Glycolysis, also called the Embden-Meyerhof pathway, is the metabolic sequence in which glucose is oxidized. It consists of nine enzymatic reactions that produce two molecules of pyruvate and two reduced equivalents of NADH, which, when introduced into the respiratory chain, will produce four molecules of ATP. When there is an absence of oxygen, glycolysis is the only pathway that produces ATP in animals. The primitive organisms originated in a world whose atmosphere lacked 02 and, therefore, glycolysis is considered the most primitive metabolic pathway. It is present in all current forms of life. It is the first part of the energy metabolism and in eukaryotic cells it occurs in the cytosol.

- In this phase, 2 ATP and 2 NADH are formed for each glucose molecule
- The overall reaction of glycolysis is:

Glucose+2 NAD+ADP+2Pi ↔ 2 NADH+2 piruvate+2 ATP+4 H+

Adenosine triphosphate or adenine triphosphate (ATP)

ATP is a molecule present in all living beings. It is the main source of energy usable by the cells to carry out their activities. It is caused by the metabolism of food in the mitochondria. The main function of the ATP is the exchange of energy and the catalytic function.

Constitution of the ATP

- Adenosine: constituted by adenine and ribose, a sugar of five carbons.
- The three phosphates that the molecule has are composed of one phosphorus atom and four oxygen atoms and the whole is bound to the ribose.
- The two bridges between the phosphate groups are high energy junctions, that is, when the enzymes break them, they yield their energy easily.
- ATP is formed in the process of food digestion when the Krebs cycle, the pentose phosphate cycle, oxidative phosphorylation, glycolysis, and the digestion of fats and proteins are carried out. Once the metabolic processes have been carried out and the ATP and other energy molecules (FAD, NADH, ADP, etc.) have formed, the deer can use them to obtain energy and transform it into physiological movements and patterns of growth, development, reproduction, production and growth of the antler.

The Krebs cycle = Tricarboxylic acid cycle = Citric acid cycle

It is a process carried out within the mitochondria of deer cells where reduced enzymes are generated that are those that possess energy. This cycle does not generate direct energy but forms highly energetic molecules that allow phosphorylating ADP to ATP. It is a succession of chemical reactions through which the final decomposition of the molecules, of the food that make up the diet of the deer, is carried out, resulting in the production of carbon dioxide, water and energy.

The krebs cycle is carried out by the action of eight enzymes:

1. Aconitase
2. Isocitrate dehydrogenase
3. Tarato dehydrogenase
4. Succinyl CoA synthetase
5. Succinate dehydrogenase
6. Fumarase
7. Malate dehydrogenase
8. Citrate synthetase.

Foods before being able to enter the citric acid cycle must be broken down into small units called Acetyl CoA as follows:

Fats → Acids → Fatty → Acetyl CoA
- Carbohydrates → Glucose → Pyruvate → Acetyl → CoA
Proteins → Amino Acids → Acetyl → CoA

The Acetyl CoA joins with an oxaloacetate molecule initiating the Ktebs cycle.

- The cell uses the ATP and GTP molecules as fuel in many processes.
- The original oxaloacetate molecule is regenerated at the end of the cycle. This molecule can then react with another acetyl group and start the cycle again.
- Energy is produced at each turn of the cycle.
- In the cycle, only acetyl and Acetyl CoA groups are destroyed.
- Both the enzymes that carry out the different reactions and the intermediate compounds on which they act can be used again and again.
- Many of the intermediate compounds that are produced in the cycle are also used as building materials for the synthesis of amino acids, carbohydrates and other cellular metabolites.
- An NADH is equivalent to three ATP.
- A GTP is equivalent to one ATP.
- An FAD is equivalent to 2 ATP

Phosphorylation or mitochondrial respiratory chain

It is a substitution reaction in which the phosphate group participates in its PO-32 form. It is important in the reaction mechanisms in which adenosine triphosphate (ATP) intervenes in the cells of living organisms. The energy obtained in respiration or in photosynthesis is used to add the third phosphate group to ADP (adenosine diphosphate) and convert it to ATP. This molecule stores that energy, which is available to the cell. The elimination of a phosphate group in the ATP supposes the release of 30.6 kJ/mol.

Gluconeogenesis

It is the synthesis of glucose from non-carbohydrate compounds. Lactic acid (lactose) stored in the muscles is transported through the blood to the liver and there undergoes a series of biochemical reactions catalyzed by enzymes that manage to turn it back into glucose, this glucose can undergo the process of glycolysis again and being converted to pyruvate, then Acetyl CoA and can thus enter the Krebs cycle. It is in this way that the lactate accumulated in the muscles serves to generate energy at the moment when the deer lacks it and is not feeding properly. If the deer does not consume the necessary amount of food, it is necessary to use their muscular reserves and therefore the loss of weight is imminent.

It is an anabolic metabolic pathway that allows the synthesis of glucose from non-carbohydrate precursors (which neither come from nor are glucose). It includes the use of several amino acids, lactate, pyruvate, glycerol and any of the intermediates of the tricarboxylic acid cycle (or Krebs cycle) as carbon sources for the pathway, all amino acids, except leucine and lysine, can supply carbon for the net synthesis of glucose. It occurs almost exclusively in the liver (10% in the kidneys). It is a very important process, since in metabolic states such as fasting, higher organisms, such as deer, can synthesize glucose from other substances. The enzymes that participate in the glycolytic pathway also participate in gluconeogenesis, they are differentiated by three irreversible reactions that use enzymes specific to this process and that condition the two metabolic detours of this pathway. These reactions are:

1. From glucose to glucose-6- (P).
2. From fructose-6- (P) to fructose-1,6-bisphosphate.
3. From phosphoenolpyruvate to pyruvic acid.

Digestion of the structural carbohydrates of plants

There is considerable variation between plant species with respect to the concentration and composition of structural carbohydrates. The concentration of cellulose is typically higher, in the walls of legumes, than those of grasses. This reflects a much lower concentration of hemicellulose in legumes compared to grasses. The concentration of cellulose often seems to be similar among grasses. However, warm climate perennial grasses have more abundant structural carbohydrates than temperate climate grasses.

It is known that approximately 50% of the organic carbon in the earth is bound to the cellulose molecule. What represents a huge source of energy, even when the cells of vertebrates do not produce the amount of cellulases necessary to break all this abundant material. However, many microbes secrete cellulases that allow them to use cellulose from diet and other plant materials. The polysaccharide cellulose as well as other carbohydrates of the plants that the deer consumes, during the ruminal fermentation, is converted into volatile fatty acids, which eventually are transported, through the ruminal wall, into the bloodstream and they are anabolized in other compounds.

The cellulolytic bacteria that are found in the rumen of the deer, of which *Ruminococcus flavefaciens, R. albus* and *Fibrobacter succionogenes* are the most important and are responsible for the cellulose digestion that occurs in the rumen. Ciliated protozoa and fungi that have also been identified in ruminal microbial populations have cellulolytic activity; however, its contribution to the degradation of cellulose is relatively minor. The cellulolytic bacteria adhere to the surface of the cell wall, placing their enzymes near the substrate. Three basic enzymatic activities are involved 1) endo-β-1,4-glucanase; which breaks at random the polysaccharide in oligosaccharides, 2) exo-β-1,4-glucanase which attacks the non-reducing end of the oligosaccharides, giving cellobiose and 3) β-1,4-glucosidase, which hydrolyzes the cellobiose in glucose. The amount in which native cellulose is used by ruminal microorganisms is limited by its association with lignin and other constituents of the cell wall

There are, however, intrinsic factors which can limit the rate at which cellulose is digested in the rumen of deer. The crystallinity of cellulose has been suggested as a factor in reducing the accessibility of cellulose to enzymatic attack. The degradation of cellulose has been shown to be inversely proportional to the degree of crystallinity for purified substrates. However, now, there is little evidence that crystallinity is a limiting factor in the degradation rate of native celluloses by ruminal microbes.

The degradation of hemicellulose in the rumen of the deer occurs in a manner analogous to that of cellulose but involves a broader arrangement of enzymatic activities. The same cellulolytic bacteria mentioned above, which are responsible for most of the degradation of cellulose in the rumen, are also the most important hemicellulolytic bacteria. In addition, *Butirivibrio fibrisolvens*, which has a relatively minor role in the degradation of cellulose, has a proportionally greater role in the degradation of xylans that are components of hemicellulose.

Some ruminal fungi and protozoa also have hemicellulolytic activity, but their activity in degrading hemicellulose is relatively minor, compared with bacteria. Isolated hemicelluloses are generally digested completely by ruminal microorganisms, degradation occurs through enzymes endo and exoglicanases, which depolymerize and solubilize the main polysaccharide chains. Substitute groups and side chains are removed from hemicellulosic polysaccharides, and subsequently degraded by activities of several glycosidases.

Composition and availability of cellular content

The cellular content in herbs (45 to 75%) and shrub plants (37 to 81%) are the main components of the diet of white-tailed deer, higher than that of grasses (18 to 28%) and is formed by soluble elements of the cell cytoplasm (proteins, lipids, carbohydrates and minerals) that are generally extracted during mastication and are completely digested during its passage through the digestive tract of deer. After proteins, carbohydrates are the most abundant in plants that form (accumulate) with surplus growth from photosynthesis. They are stored in leaves and stems; the amount of carbohydrates is affected by the temperature during growth, maturity stage, application of fertilizers and availability of nutrients in the soil. It has been reported that both tropical and temperate legumes contain high concentrations of pectins (polymers of galacturonic acid) and starch is stored mainly in leaves and stems. In contrast, sucrose, starch and small amounts of fructose are the main carbohydrates that are stored in tropical grasses. The free monosaccharides; sucrose and fructose and possibly pectins are rapidly fermented in the rumen, but starch that has been reported as relatively insoluble, could be a source of surplus nutrients for deer.

The average annual cell wall content of the forage of 32 shrub plants, consumed by the white-tailed deer in northern Mexico and the southern US, was 37% within a range of 19 to 63%. Likewise, average annual content of cellulose of shrub species was 13%, within a range of 5 to 31%. The average annual content of hemicellulose was 12%, which falls within a range of 4 to 28%). The average cellular wall content (37%) of the shrubby trees is relatively constant throughout the year and low compared with another group of plants such as native grasses, although comparable with the cell wall content of the herbs. Apparently, forage plants with low cell wall content are indicative of a high nutritional quality due to their high cellular content, which is abundant in non-structural carbohydrates, such as the starch that plants store mainly in the leaves and seeds and are very soluble. This has also been demonstrated by comparing the nutritional quality of shrub legumes that grow in tropical climates and in temperate climates and in tropical grasses.

Chapter 7

Lipids

Abstract.- In the organism of the deer, the lipids provide energy for normal maintenance and for productive functions, serve as a source of essential fatty acids, as a carrier of fat-soluble vitamins and form the lipid bilayers of the membranes where they coat organs and give consistency, they are transporters since they favor or facilitate the chemical reactions that take place in the deer and fulfill this function the lipid vitamins, the steroid hormones and the prostaglandins. The digestion and absorption of the fatty acids in the deer occurs in the rumen, while in the nonruminants the lipolysis occurs in the lower digestive tract, first in the small intestine near the absorption site. Apparently, the white-tailed deer does not have specific lipid requirements; however, fats and oils in their diets provide an important source of energy, essential fatty acids and fat-soluble vitamins (A, D, E and K). In fact, lipids have 2.5 times more energy than carbohydrates or proteins. The deer stores enough fat in its adipose tissue during summer and autumn as a reservoir for winter. Although it does not require lipids in the diet to carry out the storage because the deer converts the energy in the carbohydrates into saturated fatty acids that are stored in the adipose tissue, later the fat is used during the critical periods. This is a natural phenomenon and, one of the reasons why the food requirements and food consumption of the deer in winter are lower.

Introduction

Lipids are organic biomolecules formed basically by carbon and hydrogen and generally, to a lesser extent, also oxygen. In addition, they can occasionally also contain phosphorus, nitrogen and sulfur. It is a group of very heterogeneous substances that only have these two characteristics in common: they are insoluble in water and are soluble in organic solvents, such as ether, chloroform, benzene, etc. When lipids are in the solid state at room temperature, they are called fats. Generally, fats are composed of fatty acids with a high degree of saturation (hydrogenation). The essential fatty acids (EFA) are those that must be supplied in the diet of the deer because they cannot be metabolically produced in the speed and quantity required by the body. Even when the white-tailed deer consumes plants with low fat content, it is required to carry out important functions to maintain homeostasis in their organism. The white-tailed deer deposits fat during the summer and autumn to use it in winter. The deer converts the energy of carbohydrates into saturated fats that are deposited in adipose tissue, using it during difficult conditions. This is a natural phenomenon and represents one of the reasons why the deer has a low consumption of food in winter. Adipose tissue or storage fat is in a form that is quite available to be used to produce energy when required and the marbled fat in the deer muscle is low.

Important lipids for white-tailed deer are the following:

Phospholipids

Phospholipids have their greatest importance as constituents of the lipoprotein complexes of biological membranes. Its distribution is very wide, being particularly abundant in the heart, kidney and nervous system examples are:.

- Lecithins (choline) are, like fats, esters of glycerin. The main acids that form them are palmitic, arachidonic, oleic and stearic. Lecithins are waxy-looking solids that, exposed to air, turn brown rapidly due to oxidation.
- Cephalins are differentiated from lecithins by having the colamine as a nitrogenous base.
- Sphingomyelins do not contain glycerin and are composed of fatty acids, phosphoric acid, choline and sphingosine. Sphingomyelins are found mainly in nerve tissues.

Galactolipids

Most of the fat in the diet of ruminants comes from forage plants with crude fat contents of around 4%, an important portion comes from galactolipids. 95% of the fatty acids present correspond to linoleic. The rumen microorganisms separate the galactolipids to give galactose, fatty acids and glycerol.

Cerebrósidos

They are compounds that exist mostly in the nervous tissue. They are formed by a fatty acid, normally of high molecular weight, attached to the amino group of sphingosines, which in turn carries an alcoholic group esterified with a molecule of hexose, usually with galactose and less frequently with glucose.

Waxes

The waxes are simple lipids formed by the combination of a fatty acid with a monoalcohol of high molecular weight, usually they are solid at room temperature. The waxes often have a protective mission, they are not hydrolyzed easily and lack nutritional value.

Steroids

Among these are compounds such as sterols, bile acids, adrenal hormones and sex hormones. Sterols can be divided into:

- Phytosterols, plant origin.
- Micosteroles, fungal origin.
- Zoosterol, animal origin.

Cholesterol is a zoosterol that has an important representation in the brain and can be synthesized by the deer organism. 7-dehydrocholesterol, which is derived from cholesterol, is a precursor of vitamin D3, in which it is converted by the action of UV light. Ergosterol is a phytosterol, it is a precursor of ergocalciferol or vitamin D2, in which it is converted by UV action.

Bile acids: They are important in the duodenum, where they help to emulsify fats and activate the lipase. However, the white tail deer does not have a gall bladder so it does not accumulate bile in its organism.

Adrenal hormones: Among them we have corticosterone and cortisol and control the production and utilization of glucose and the mobilization and production of fats.

Terpenes

To the group of the terpenes belong substances of as much biological importance as the pigments carotene and lycopene, vitamin A, the rest phytol of chlorophyll and several of the so-called essential oils.

Classification according to its importance in nutrition

The lipids carry out important biochemical and physiological functions in the tissues of the deer and can be classified as follows:

Simple lipids.- are esters of fatty acids with several alcohols; for example: fats, oils and waxes

Compound lipids.- are esters of fatty acids that contain other groups, in addition to an alcohol and a fatty acid. For example: phospholipids, glycolipids and lipoproteins.

Derived lipids.- are those substances that are derived from the groups mentioned above through hydrolysis. For example: free fatty acids.

Sterols.- are lipids with complex structures with rings similar to those of phenanthrene, of which the precursor of most of them is cholesterol. For example: vitamin D, hormones of the adrenal gland (cortisol, cortisone, aldosterone).

Terpenes.- have a structure of the isoprene class, which is a precursor to many important lipids in animal nutrition, and they all have carbons in multiples of five. For example: vitamin A (20 carbons) and vitamin K (40 carbons).

When the phospholipids are subjected to hydrolysis, fatty acids, phosphoric acid and generally glycerol and a nitrogenous base are obtained. All the phospholipids that contain the vitamin choline are called lecithins and the phospholipids that do not contain it are called cephalins. The first are the most important in the nutrition of the deer, while the second are generally located in the nervous tissue.

Lipid functions

The functions of lipids in deer, in general can be classified as follows:

1. Supply energy for normal maintenance and for productive functions
2. Serve as a source of essential fatty acids
3. As a carrier of fat-soluble vitamins
4. They form the lipid bilayers of the membranes. Coating organs and give consistency
5. They are transporters since they favor or facilitate the chemical reactions that take place in body of deer. This function is fulfilled by lipid vitamins, steroid hormones and prostaglandins.

Generally, carbohydrates provide all the energy needed by deer, except for that which provides essential fatty acids. The absorption of liposoluble vitamins (A, D, E and K) is a function of the digestion and absorption of fats in the digestive tract of deer.

The lipids in the body of the white-tailed deer, in their majority, are in the form of fats, even though the lipids in their diets are mainly composed of unsaturated fatty acids. This phenomenon is since most of the fatty acids are saturated with hydrogen in the digestive tract due to the high reductive potential (hydrogen) of the rumen. The main component of adipose tissue of white-tailed deer is stearic acid, which is an 18-carbon fatty acid completely saturated with hydrogen. The unsaturated fatty acids: oleic, linoleic and linolenic contain 18 C with one, two and three double bonds, respectively, are the main components of the plants consumed by white-tailed deer and are the main precursors of stearic acid. On the other hand, when the lipids are in liquid form at room temperature they are called oils.

Lipids on plants

From a nutritional point of view, lipids can be grouped into compounds stored in seeds (mainly triglycerides), lipids in leaves (galactolipids and phospholipids), and a miscellaneous assortment of waxes, carotenoids, chlorophylls, essential oils and other ether-soluble substances. Some of these substances, particularly essential oils may have anti-cellulite activity in the rumen of deer. Essential oils are abundant in the native shrub plants consumed by white-tailed deer in northeastern Mexico and are made up of a diverse group of organic substances from plants that have common properties such as volatility and solubility in organic solvents. Esters, ethers, phenols and members of the family of the cyanimic acid that is related to the lignins are essential oils. Some herbivorous browsing animals such as white-tailed deer can adapt to consume these substances and detoxify by excreting essential oils in the urine.

Triglycerides are usually confined to the seed of the plant. Consequently, these are the main component of the lipids of concentrated foods, but they are insignificant in forages. The leaf lipids are mainly galactolipids involving glycerol, galactose and unsaturated fatty acids with sulfonated groups. Leaf lipids are generally more polar than triglycerides and have lower energy values than those that can be estimated by the factor 2.25 used to calculate total digestible nutrients (TDN). The content of galactolipids and triglycerides is a good estimate of the total lipids of the food used by the deer.

The fatty acids associated with galactolipids and some of the triglycerides of the seed organs are relatively unsaturated and contain high amounts of linoleic and linolenic acid. The galactolipids, a metabolically important fraction in the plant are probably less variable in composition than the triglycerides of the seed. These concentrations decline with the age of the plant and vary with the proportion of leaves: stems and other metabolically inactive

Fatty acids are the main components of lipids and make up the most important reserve of energy stored in deer. However, deer diets are normally very low in lipids because the plants that are their source of food contain small amounts; from 4 to 6% made up of 1.5 to 4% of simple lipids, 0.5 to 1% of waxes, 0.5 to 1% of sterols and 0.5 to 1% of phospholipids containing salts of phosphatidic acid. These characteristics of the diet may require both metabolic adaptations and methods to conserve essential fatty acids. The rumen is intolerant to high levels of fat that can alter fermentation. This situation in the operation of ruminants contrasts with newborn ruminants who ingest milk and about 30% or more of the dry matter consumed is fat, representing 50% or more of their caloric intake. Other diverse adaptations with respect to lipids are probably involved among ruminant groups, for example, African antelopes have very high vitamin E requirements compared to domestic ruminants. Many similar specializations can occur with the lipid material of overflow or escape from the rumen.

In several metabolic systems that involve fatty acids, they are derived from glucose. Glucose from the diet is low in the metabolism of ruminants; However, ruminants may involve mechanisms for their conservation, the most important of all is that related to the lack of metabolic pathways for the conversion of glucose to fatty acids. About 90% of the fat synthesis in ruminants occurs in adipose tissue. The liver, which is the main lipogenic site in many non-ruminant species (pigs, humans, rats, mice, chickens, etc.) contributes only 5% in ruminants.

Microbial lipids

Biosynthetic modification of lipids by deer rumen bacteria involves the formation of odd and branched-chain fatty acids, probably by incorporation of propionyl, 2-methylbutyryl and 3-methylbutyryl residues into the carbon skeleton. Some bacteria (*Fibrobacter succinogenes, Treponema* and *Selenomonas*) require n-valeric acid for the synthesis of odd-chain fatty acids. Propionate can also be incorporated into fats through the metabolism of deer, especially when propionate is produced in large quantities. Branched-chain fatty acids contribute to the lower melting point of membrane lipids in the absence of significant amounts of polyunsaturated fatty acids. The composition of fatty acids of microbes is closely linked to the metabolism of lipids in the rumen, since microbial lipids are the result of modifications to the lipids of the diet and to the products of the synthesis. The fatty acids coming from the hydrolysis of triglycerides are precipitated as calcium salts that become part of the microbes.

While the normally found amounts of unsaturated lipids in forages do not affect fermentation in the rumen, an excess of these can cause the suppression of cellulolytic and methanogenic bacteria and, in general, all gram-negative bacteria. The mode of supply of this type of lipids determines the magnitude and direction of the changes produced. The supply in small frequent doses gives time to the hydrogenation of the lipids before any inhibition occurs. The inhibition of methanogenesis reduces the synthesis of milk by the mammary gland to favor the formation of propionate in the rumen and the metabolic changes in the deer linked to this type of fermentation. On the contrary, unsaturated fatty acids protected or supplied in small and frequent doses, increase milk fat.

Essential fatty acids

The essential fatty acids (EFA) are those that have to be supplied in the diet of the deer because they cannot be metabolically produced in the speed and quantity required by your body. The essential fatty acids are: 1) linoleic acid (18 carbons and 2 double bonds), 2) linolenic acid (18 carbons and 3 double bonds) and 3) arachidonic acid (20 carbons and 4 double bonds). Apparently, the deer has a lower EFA requirement than non-ruminants. Fawns are born with lower reserves of polyunsaturated acids, such as essential fatty acids, than non-ruminants, increasing them after birth under normal feeding conditions. The adult deer makes more efficient use of essential fatty acids by selectively incorporating them into cholesterol and phospholipid esters as a mechanism of adaptation to rumen hydrogenation.

An interesting point is that sufficient quantities of unsaturated fatty acids (linoleic and linolenic) can escape from the hydrogenation in the rumen under normal feeding conditions to satisfy the EFA of deer requirements. These same acids are necessary precursors in the production of prostaglandins, substances that resemble hormones which regulate the essential activities of the cell and determine the health status of all cells, tissues and organs.

Linoleic acid.- It is an essential unsaturated fatty acid, more specifically polyunsaturated (two double bonds) and belonging to the omega-6 group since the first double bond is after carbon 6, counting from the methyl end (-CH3).

Linolenic acid.- It is an essential fatty acid omega-3 (the α isomer) or omega 6 (the γ isomer), formed by a chain of 18 carbons with three double bonds at positions 9, 12 and 15.

Arachidonic acid or eicosatetraenoic acid.- It is a polyunsaturated essential fatty acid of the omega-6 series, formed by a chain of 20 carbons with four cis double bonds at positions 5, 8, 11 and 14, which is why it is the acid 20: 4 (5,8,11 , 14).

Lipids of animal origin

The fats of ruminants such as deer are characteristically hard (saturated) compared to those of non-ruminants. It is possible to observe the considerable capacity of the rumen of deer to change and hydrogenate unsaturated fatty acids. Another peculiarity of ruminants is the contrasting comparison of milk fat compared to storage fat. Short chain lipids are present in milk fat but are almost absent in other ruminant lipids and milk fats from non-ruminants. The quality and composition of milk fat are associated with the metabolism of propionate and carbohydrates in the body of deer.

The lipids in the deer organism are extraordinary because they contain extra carbon and branched fatty acids, which reflect the absorption and incorporation of microbial lipids and some modified components of the plant (phytanic acid) into the fat of deer. Ruminant tissues can dehydrogenate and change the length of fatty acid chains (oleic acid is apparently derived from palmitic acid through this process).

Lipids from the diet of all non-ruminant herbivores (without pre-gastric fermentation) are unsaturated. Dietary lipids consumed by omnivores and carnivores may be more variable, depending on the source of the diet, but ruminant fats have greater difficulty in changing in the diet, because of saturation during ruminal fermentation. The fats deposited in ruminants are not less alterable than those of nonruminant species; as long as during ruminal fermentation they were not fermented.

Ruminal metabolism of lipids

The digestion and absorption of the fatty acids in the deer occurs in the rumen, while in the nonruminants the lipolysis occurs in the lower digestive tract, first in the small intestine near the absorption site. In both types of animals, the long chain fatty acids are absorbed via the lymphatic system. The fatty acids are neutralized at ruminal pH and pass as soaps (micelles). Potassium soaps are easily absorbed in the small intestine of the deer. There is evidence that calcium soaps (much less soluble) can escape absorption and appear in the feces of animals. Saturated fatty acids are absorbed in a smaller proportion than unsaturated ones. The availability for absorption decreases as the length of the acid chain increases. Even so, ruminants ordinarily absorb more fatty acids, with a true digestibility close to 100%. Approximately 15% of the lipids absorbed in the duodenum of the deer consist mainly of bacterial phospholipids because the rumen microorganisms synthesize fatty acids, most of which are incorporated into the phospholipids that form the cell membrane.

Synthesis of milk fat

The mammary gland during lactation is the site of the highest synthesis of triglycerides. Fat represents more than 50% of the calories in milk. The milk fat of ruminants such as deer comes from adipose tissue. The main peculiarity is the content of short chain fatty acids, which are absent in many other fats. The synthesis of shorter chains is the main lipogenic activity of the mammary gland. The source of carbon for the synthesis of fatty acids in milk fat depends on the length of the chain. The elongation of the capric acid chain is carried out by sequentially joining two to two carbon atoms. The contribution of acetate and β-hydroxybutyrate (BHBA) to long chains is very limited. The mammary gland uses BHBA along with the acetate to add the carbon for the short chain fatty acids in the milk fat. The enzyme synthetase, for fatty acids of this gland, prefers first butyryl coenzyme A, in comparison with the adipose tissue that uses acetyl coenzyme A for this purpose.

Requirements of deer lipids

Apparently, the white-tailed deer does not have specific lipid requirements; however, fats and oils in their diets provide an important source of energy, essential fatty acids and fat-soluble vitamins (A, D, E and K). In fact, lipids have 2.5 times more energy than carbohydrates or proteins. Therefore, acorns and wild nuts, which have a high content of lipids, are important sources of energy. Deer milk contains 7.7% fat, almost double that of bovine milk.

The deer stores enough fat in its adipose tissue during summer and autumn as a reservoir for winter. Although it does not require lipids in the diet to carry out the storage because the deer converts the energy in the carbohydrates into saturated fatty acids that are stored in the adipose tissue, later the fat is used during the critical periods. This is a natural phenomenon and, one of the reasons why the food requirements and food consumption of the deer in winter are low. Fat in adipose tissue is mobilized very easily by hormonal mechanisms, to be used to produce energy when required.

The Table 7.1. shows the levels of fat, cholesterol and calories from the meat of different animal species. Poultry meat contains low levels of fat and cholesterol, followed by meat from wild animals, such as white-tailed deer. However, the calorie content is very similar between animals. On the other hand, the cardiac muscle contains 275 mg of cholesterol per 100 g of muscle and the liver contains 450 mg in the same portion.

Table 7.1. Levels of fat, cholesterol and calories from the meat of different animal species. (100 g) of meat of wild animals, cattle, pig and birds without skin

Meats	Fat, g	Cholesterol, mg	Calories
Cattle	2.7	69	158
Pigs	4.9	71	165
White-tailed deer	1.4	113	153
Mule deer	1.6	85	151
Antelope	1.0	113	148
Buffalo	3.2	45	146
Chipmunk	3.2	83	149
Rabbit	2.4	77	144
Chicken	0.7	58	140
Turkey	1.5	60	146
Wild turkey	1.1	58	158

Chapter 8

Energy

> **Abstract.-** Energy is not really a food for the deer, rather it is a characteristic of food. Protein, lipids and carbohydrates contain energy, while water, vitamins and minerals do not. It has been determined that the carbohydrates upon oxidation produce 4.1 kcal/g of carbohydrate. The proteins produce 5.65 kcal/g of protein and the lipids produce 9.45 kcal/g of lipid. Therefore, lipids produce 2.25 times more energy in the body than carbohydrates. The volatile fatty acids are the main source of energy for the white-tailed deer; They provide 66 to 75% of the energy available for the metabolism. The main AGV in order of abundance are acetic, propionic, butyric, isobutyric, valeric and isovaleric. The production of AGV in the rumen of the deer is markedly influenced by the diet and the type of methanogenic population in the rumen. The deer has evolved adapting to the efficiency of gluconeogenesis, while the lower digestive tract has been conditioned to the lack of good quality carbohydrates such as starch. Apparently, the native forbs that grow in northeastern Mexico and that are consumed by the white-tailed deer have, on average, a higher content of ME followed by the shrubs and at the end of the grasses. Except for the juveniles of 4 to 6 months and 7 to 11 months of age, the forbs contain enough ME to cover the needs of the deer in any physiological state. Shrubs satisfy the demand for ME only for maintenance. Grasses contain marginally low amounts for all physiological states. There is not any plant that can maintain, throughout the year, the levels of energy required for deer for a good growth and development.

Introduction

Energy can be defined as the ability of a body to perform a job. Plants obtain their energy directly from sunlight, while deer must obtain a constant supply of energy through their food. The white-tailed deer requires energy to maintain its bodily functions: move, grow, produce milk, reproduce and develop antlers in males. They get their energy mainly from carbohydrates (sugar, starch and cellulose) and to a lesser extent from fats in the diet. The microorganisms in the rumen of the deer transform carbohydrates, mainly cellulose (which cannot be digested by nonruminants), into volatile fatty acids (acetate, propionate, butyrate, isobutyrate, isovalerate and 2-

methylbutyrate), which are simpler molecules than they can satisfy most of the needs of energy for deer. Other carbohydrates (sugars and starches) are also used as an energy source.

Types of energy

Energy is not really a food for the deer, rather it is a characteristic of food. Protein, lipids and carbohydrates contain energy, while water, vitamins and minerals do not. Energy is usually expressed in terms of calories (c), or kilocalories (Kcal = 1000 calories). The joule (J), which is precisely defined with respect to certain electrical measurements. Because joule is a very small unit, it is more common to find megajoule (MJ, 1,000,000 J) or kilojoule (kJ, 1,000 J) in animal nutrition publications. Another unit that is commonly seen, especially in older publications, is the calorie, which is equal to 4. A Joule equals 4,184 cal. In some circumstances, it is preferred to use the system of total digestible nutrients (NDT) where the energy is expressed as a percentage of the diet or kg per day. However, energy is probably the most variable of the nutrients required by the deer because it is strongly affected by the environment.

Deer heat production does not relate very well to body weight, but rather relates to body surface or volume. The previous thing is verified when raising to the 0.75 the corporal weight of the animal and it is arrived at a concept that is called metabolic weight; which is the body weight at 0.75 ($BW^{0.75}$).

It has been determined that the carbohydrates upon oxidation produce 4.1 kcal/g of carbohydrate. The proteins produce 5.65 kcal/g of protein and the lipids produce 9.45 kcal/g of lipid. Therefore, lipids produce 2.25 times more energy in the body than carbohydrates. This is because the chemical energy varies inversely with the carbon: hydrogen ratio and the oxygen and nitrogen content. The differences between these nutrients mainly reflect the oxidation state of the initial compound. The more oxygen required to oxidize a compound, the more energy it produces; for example, glucose has an empirical formula of $C6H12O6$, that is, an oxygen atom for each carbon atom, while a fat molecule has 6 oxygen atoms and 57 carbon atoms; therefore, the fat needs more oxygen to carry out the oxidation and releases much more heat during the procedure.

Fractions on how energy is divided in terms of use by animals

The energy contained in the plants that the deer consumes is digested, metabolized and used to carry out physiological functions and maintain homeostasis. Depending on the type of nutrients contained in the plants and the secondary compounds they contain, determines the percentage of energy use by the body. Energy losses occur in excretory products such as feces, urine, gases and energy that is lost during fermentation in the rumen (caloric increase). For example, the average energy that is lost when the white-tailed deer consumes foliage of shrubs that grow in northeastern Mexico and southern Texas, USA, corresponds to 47%, while the loss with the forbs consumption of the same region is 43% and the one that is lost when the deer consumes native pastures is 61%.

Volatile fatty acids (VFA)

The AGV are the main source of energy for the white-tailed deer; They provide 66 to 75% of the energy available for your metabolism. VFA are a waste product of anaerobic microbial metabolism in the rumen, largely due to the fermentation of the carbohydrates in the diet, the latter providing the deer with the largest source of metabolizable energy, since only a small proportion of the Carbohydrates escape fermentation in the rumen. The removal of VFA from the rumen is vital for the maintenance of the rumen environment and for the continued growth of cellulolytic organisms.

The main AGV in order of abundance are acetic, propionic, butyric, isobutyric, valeric and isovaleric. The production of AGV in the rumen of the deer is markedly influenced by the diet and the type of methanogenic population in the rumen. Protozoa can also contribute significantly to the balance. Other acids may appear as products of the fermentation of carbohydrates, for example, lactic acid is important when starch is part of the diet which is eventually fermented to acetate, propionate and butyrate. AGV concentrations in the rumen are regulated by a balance between production and absorption. The concentration of AGV and the pH of the rumen vary because the degree of production varies because of the feeding regulations.

The relative proportions of VFA vary with diet. Acetic acid predominates in most conditions, but substantial amounts of propionic and butyric acids are always found. The AGV concentrations of the rumen are highly related to the amount absorbed. While the quantitative relationship depends on the size and ruminal capacity, as well as the rate of passage. For example, diets based on concentrates result in high concentrations of VFA compared to the amount absorbed. A small ruminal volume and a reduced rumen capacity, as in the case of deer, makes the rate of passage fast.

Volatile fatty acids are absorbed through the rumen wall in free form, and not by active transport. In the ruminal epithelium of the deer a considerable metabolism of fatty acids is generated, leading to a differential decline in concentration and more rapid absorption. At the normal pH of the rumen, only small amounts of AGV are in the form of free acids. In ruminants unlike non-ruminants, AGV and little or no glucose are absorbed.

The mechanism of absorption of the VFA is as follows: The intracellular pH of the ruminal wall and blood is more alkaline than that of the rumen, which favors the movements of free acids into the blood through the free energy of the neutralization. This condition discourages the fluidity of the VFA in the form of anions. The mechanism of absorption of AGV in the lower digestive tract of the ruminant and nonruminant is similar to that of the rumen. There could be a considerable metabolism of butyric acid in the ruminal wall leading to a differential decline in concentration, accelerating absorption. At a normal pH, only small amounts of AGV are in the form of free acids. In any case, the absorption of free acids is balanced by the formation of more free acids through the reversion of the ionization equilibrium by the action of mass. The presence of high concentrations of free acids is favored by low pH and high concentrations.

The intermediary metabolism in ruminants, such as deer, is basically the same in mammals, but differs mainly in the amounts of carbon through certain metabolic pathways, which are low with respect to glucose uptake and high uptake. acetic, propionic and butyric acids in the gastrointestinal tract. In general, acetate and butyrate are the major sources of energy (by oxidation) while propionic is reserved for gluconeogenesis. As a result, the deer has evolved adapting to the efficiency of gluconeogenesis, while the lower digestive tract has been conditioned to the lack of good quality carbohydrates such as starch. Acetate is the most absorbed component (90%) from the reticulum-rumen, besides being the most important lipogenic precursor. The VFA are absorbed in free form, in their metabolism they pass through the ruminal wall towards the portal blood, passing as neutralized anions in the blood pH. The rumen epithelium metabolizes the VFA considerably, resulting in the amounts being greater for the butyric and lower for the acetic. The portal blood circulates through the liver which takes almost all the propionate and butyrate, so that all the acetate represents 90% or more of the AGV found in the peripheral circulation .

Energy requirements

Deer requires energy for basal metabolism, which is necessary to maintain body temperature in a normal environment, maintain breathing and voluntary activity. Any activity of the individual represents a "cost" in terms of energy. The animals respond behaviorally to the thermal regime, altering the balance between heat loss and heat production, through changes in orientation, posture, activity or selection of cover for protection. The energy requirements can be calculated for the different activities, and the sum will be the total daily energy requirement. Additionally, the actual energy needs of deer are approximately twice those of maintenance. There are of course additional energy needs for growth, reproduction, gestation, lactation and antler growth in adult males. Also, there are additional energy requirements for daily activity (walking, browsing, avoiding predators, or getting away from hunters).

For females of white-tailed deer, the last third of gestation (which in this case coincides with the dry season) is the one with the highest energy metabolism. The energy requirement during this stage is 30 to 45% higher than maintenance needs, and so energy requirements during lactation can reach 2.3 times the basal metabolic rate. While for the male deer, it is the mating season that implies an increase in energy expenditure, since they must defend the females in estrus, against other males, to ensure their offspring.

However, the energy needs of any animal will depend on the intrinsic conditions of the individual (physiological state, sex, age) and extrinsic conditions such as the type of vegetation, quantity and quality of the habitat and the characteristics of temperature, humidity and precipitation of the environment. It has been determined that the lowest metabolism in white-tailed deer occurs in winter and the highest in summer, the former being an adaptation to conserve energy; that is, the need for food is lower when the available food is reduced. Although it was examined the metabolic rate of captive white-tailed deer females found no significant differences between stations, noting that perhaps the changes found in other studies, with an increase in the metabolic rate in summer, could be due to the measurement of metabolic rate out of thermoneutral zone.

The behavior is important because voluntarily the deer in general, can avoid unpleasant environmental circumstances, such as intense heat or cold and winds. Therefore, the effect of different environmental conditions (temperature, precipitation) on the thermoregulation of the animal, and on the availability and quality of the forage, can influence the metabolic requirements and energy requirements to maintain body mass. The lower maintenance requirements can be an adaptive response to low productivity environments. The lowest metabolic rates are found in species adapted to deserts that subsist on forages of low quality. When it was compared the energy expenditure between sexes in arid zones, showed that if they are different since the males spent more energy (1726±38 kcal/day/individual) than the females (1556±35 kcal/day/individual). In addition, it was reported that it is lower in arid zones compared to more northern areas, as has been pointed out by several authors.

Apparently, the native grasses that grow in northeastern Mexico and that are consumed by the white-tailed deer have, on average, a higher content of ME followed by the shrubs and at the end of the pastures. The highest values in summer and autumn may be due to higher precipitation and temperature in these seasons. Except for the juveniles of 4 to 6 months and 7 to 11 months of age, the forbs contain enough ME to cover the needs of the deer in any physiological state. Shrubs satisfy the demand for EM only for maintenance. Grasses contain marginally low amounts for all physiological states.

Forbs represent the largest source of energy for the white-tailed deer; however, they are only present in the pasture during wet seasons, especially in summer and autumn. On the other hand, shrubs by the amount of energy they provide when consumed by the deer can be placed in an intermediate term. However, in regions where winters are generally not very cold, most of the plants that are browsed by the deer remain green and have a nutrient content, including energy, very similar for most of the year, so they could be considered as a permanent source of energy, to meet the demand for energy throughout the year. Although native grasses do not provide enough energy, deer either way includes them in their diets, especially the more tender parts of the plant that contain the most available nutrients and fiber sources.

Due to its reduced metabolism when the winter is very cold, the deer is not as well adapted to the sudden deterioration of climatic conditions as is the case of cattle or sheep. The activity of the thyroid gland (which regulates the rate of basal metabolism) decreases during cold weather. The above creates a physiological state like hibernation, but of course, much less extreme. The activity is dramatically reduced compared to summer or autumn. When seeking shelter to protect themselves from the cold, it reduces its forage activity, but at the same time, it conserves its reserves of energy to maintain its body temperature. The reduction in the voluntary consumption of food is approximately 40 to 50% with respect to the consumption of food during the spring. The net energy required for maintenance is lower in January. The increase in energy demands with physical disturbances or lack of shelter in winter are potentially detrimental to the deer. Wind and humidity can also have a detrimental effect and increase energy needs

An important aspect about energy needs of deer is that they are directly related to body weight. That is, as the deer is larger, it needs less energy per unit of body weight. This is also reflected in the consumption patterns of white-tailed deer. The bigger your body is, the lower is your consumption per unit of body weight. The most important is that energy needs, and food consumption vary seasonally. Males and females have their highest consumption at the end of summer and early fall (which is when forbs with the highest nutrient content are present). This may be the period (late summer and early fall) most critical for the consumption of food for the deer because the antlers of the deer begin their growth and requires depositing sufficient reserves of fat for the run in winter, the females are lactating or weaning their fawns, and fawns are changing from a liquid diet based on milk to solid food. Once the winter begins and the mating season begins, both males and females reduce their food intake because they put all their attention on the run although during the low winter temperatures they require more energy which they obtain from the accumulated fat stores past year. Adult deer can easily lose 15 to 30 percent of their body weight in winter, although they can recover it during the spring and summer.

Deer require dramatically more energy to maintain their body temperature during the winter, especially if forced to forage or to avoid danger and maintain homeostasis in different physiological states. Energy deficits in deer, however, can result in growth retardation, weight loss, reproductive failure and ruminal dysfunction. An energy deficiency in pregnant and lactating females can result in low birth weights and premature death of fawns that depend on breast milk for their development. Adult males can grow with very small antlers if they suffered energy deficiencies during their growth.

The white tail deer is a very selective browsing and, therefore, consumes plants with high nutritional value. However, the availability of nutrients to be digested varies considerably between seasons (due to climatic changes) and between plants due to the different concentrations of secondary compounds (tannins, phenols, essential oils, etc.) that at some point are synthesized, for its own defense against herbivores, and that can accumulate in the foliage with pathological effects, which limit the voluntary consumption of food and digestibility of nutrients and, in extreme cases, cause death. Therefore, there is perhaps no plant that can maintain, throughout the year, the levels of nutrients required by the deer for proper growth and reproduction.

Chapter 9

Minerals

Abstract.- The minerals in the organism of the deer carry out diverse functions like structural components of the skeleton, (Ca, P and Mg), take part in the balance acid-base (pH) of the fluids of the organism (Na, K and Cl), they participate in enzymatic systems as activators (Zn and Cu) and an important number of them have more than one function. However, the alteration in the concentration of minerals directly affects the health of the deer and therefore their productivity, this is mainly since the food they eat has deficiencies or excesses. Mineral deficiencies can lead to tegumentary disorders (skin), non-infectious abortions, diarrhea, anemia, weight loss, loss of appetite, bone abnormalities, tetany, low fertility, and debilitating diseases, among other clinical signs. The main source of minerals is food. However, under certain circumstances, water may contribute significant amounts of minerals. Adult males store minerals in their skeleton and transfer them to the antlers if required, which may help explain why studies of mineral requirements do not match. In fact, while antlers are mineralizing, males suffer from osteoporosis (loss of minerals in the bones). Once the antlers harden, the minerals, such as Ca, lost from the bones are replaced by those from the diet of the deer. Na and P are the most commonly deficient minerals in temperate climate ecosystems. Herbs are the group of plants with the highest content of minerals, followed by shrubs, grasses represent a poor source of minerals for white-tailed deer. Cacti, flowers and fruit are important sources of all essential minerals.

Introduction

Minerals are required for the normal functioning of all the essential biochemical processes of the organism. An essential mineral can be defined as that which is required by the deer to support adequate growth, reproduction and health through its normal life cycle, when all other nutrients are in optimal amounts. Based on the identification of one or more metabolic functions, at least 15 minerals (N, P, S, K, Ca, Mg, Na, Cl, Fe, Mn, Cu, Co, I,

Mo and Se) can be classified as essentials The deficiency of each mineral in the deer results in abnormalities that can only be corrected by the deficient ore supply. The severity of the deficiency will determine the degree and type of abnormality observed. The metabolism and nutrition of minerals is similar in all animal species and, therefore, the observations made of a species can be extrapolated to other species. The mineral elements are divided into two groups according to the abundance of them in the organism: macrominerals that are elements that are found abundantly in the organism and have functions of structural and micromineral type or trace elements that are found in the organism in very small quantities and usually carry out functions as enzymatic cofactors.

Importance of minerals in deer

In the organism of the deer carry out diverse functions:

1) As structural components of the skeleton, (e.g. Ca, P and Mg).

2) Intervene in the acid-base balance (pH) of the body's fluids (e.g. Na, K and Cl).

3) Involved in enzymatic systems as activators (e.g. Zn and Cu).

4) A significant number of them have more than one function.

However, the alteration in the concentration of minerals directly affects the health of the deer and therefore their productivity, this is mainly since the food they eat has deficiencies or excesses. Mineral deficiencies can cause tegumentary disorders (skin), non-infectious abortions, diarrhea, anemia, weight loss, loss of appetite, bone abnormalities, tetany, low fertility, and extenuation diseases, among other clinical signs. And, on the other hand, all mineral elements, whether indispensable or not indispensable, can affect the deer in an adverse way, if consumed at excessively high levels.

In ruminal fermentation the minerals Ca, K, Fe, Zn and Mn are of great importance for the growth of ruminal bacteria and, therefore, in the digestion of the organic matter carried out by the deer (Table 9.1). In addition, P is of utmost importance for the proper metabolism and health of the ruminal microflora. P is part of the nucleic acids (DNA and RNA) found in all bacterial cells. In the rumen bacterial cells 10.3% of DNA and 9.6% of RNA are constituted by this mineral. Most of the RNA from the cells is found in the ribosome and, the ribosomal content in the bacteria is directly related to bacterial growth, and therefore, cellulolytic activity. Zn is essential for all living biological systems. The lack of availability of Zn for bacteria inhibits the multiplication of these, in addition to affecting the ability of the cellulolytic bacteria to adhere to the cell wall of plant tissue and actively exercise their cellulolytic capacity. In contrast, Mn is required for the growth of most cells and to play an important role in the decarboxylation reactions of the tricarboxylic acid cycle. In addition, it has been shown to stimulate the binding of CO_2 in the production of succinic acid by ruminal bacteria. Therefore, in the deer two types of requirements for these minerals must be currently considered: one for the animal and another for the microbes of the rumen.

Table 9.1. Functions of minerals in conjunction with ruminal microorganisms.

Minerals	Functions
P	Energetic processes and cell reproduction
Mg, Fe, Zn, Cu y Mb	They are activators of microbial enzymes
Co	Production of vitamin cyanocobalamin (B12)
S	Digestion of cellulose, assimilation of non-protein nitrogen (NNP) and synthesis of amino acids and vitamins of the B complex
Na, Cl y K	Metabolic processes, pH regulation and osmotic pressure

Availability of minerals in forages

The main source of minerals is food. However, under certain circumstances, water can contribute significant amounts of the following minerals: iodine, manganese, iron, sulfur, sodium, chlorine and magnesium. Different management practices (grazing pressure) and flooding can cause a substantial amount of soil that can provide significant amounts of cobalt, iron, manganese or selenium to grazing animals. For cervids, such as deer, the soil, in certain places, can be an important source of sodium, especially during the spring. Dry feces are almost pure clay and it is assumed that wild animals consume them to cure diarrhea.

The best habitat for the deer to develop optimally, is one that contains all the essential elements; as they are the vegetal cover, water and space that it has, but the most important one is the availability and quality of the food. Habitat conditions influence the size of the population and the physical appearance and size of the antlers. The content of minerals in the forages consumed by the deer changes with the season of the year; because the availability of nutrients in the soil and the ability of the root system to absorb them are affected by weather patterns. This variability can produce situations in which the food fills the deer's mineral requirements for a few months but fails the rest of the year. Generally, important nutrients such as minerals go parallel to the digestibility of a given forage.

Comparatively, little is known about the true availability of mineral elements in plants. It is speculated, however, that the content of fiber and lignin can promote fecal losses of magnesium, zinc and iron, either through their union via exchange of cations or by the presence of forms not available in the matrix of the fiber. The cell walls of the forage contain small amounts of iron and zinc that are not interchangeable. The silica fraction can be responsible for binding with these two minerals. A theoretical explanation of the siliceous inhibition of cellulolytic digestion postulates that most of the silica consumed can create deficiencies of trace minerals in the rumen bacteria.

Macrominerals

Seven are the main macrominerals that are required by the deer to carry out their vital functions. The functions and symptoms of deficiency are described in Table 9.2.

Table 9.2. Funciones y síntomas de deficiencia de los macrominerales esenciales en el organismo del venado

Mineral	Functions	Symptoms of deficiency
Calcium	- It is a structural component of the skeleton. - Controls the excitability of nerves and muscles. - It is necessary for blood coagulation. - It intervenes in the ionic movement of sodium and potassium, restricting the movement of potassium.	• Decrease in milk production. • Muscle weakness and rumen and heart dysfunction. • Rickets in young animals. • Osteomalacia in adult animals. • Fibrous osteodystrophy. • Tetanus hypocalcemia.
Phosphorous	- Intervenes as a component of the skeleton. - It is a component of phospholipids (lecithins). - Acts in the energy metabolism as a component of ATP, ADP and AMP. - It is part of the nucleic acids RNA and DNA. - It is a constituent of several enzyme systems.	• Decrease in food consumption. • Low growth. • Decrease in milk production. • Failures in reproduction and lethargy. • Rickets in young animals. • Osteomalacia in adult animals. • Appetite depraved or pica.
Magnesium	-It is required for a normal skeletal development, as a constituent of the bones. -It is necessary in the oxidative phosphorylation of atp in the mitochondria of the cardiac muscle. -It is needed for the activation of enzymes and enzymatic reactions in which atp intervenes (e.g. muscle contraction, synthesis of proteins, nucleic acids, fats and coenzymes and in the use of glucose).	-Tetanus hypomagnesemia. -Decreased activity of magnesium-dependent enzymes.
Potassium	-It intervenes in the osmotic balance. -In the acid base equilibrium (pH). -It is required for several enzymatic reactions. -It facilitates the cellular uptake of neutral amino acids. -It influences the metabolism of glucose.	-It is associated with abnormal electrocardiograms in calves, chickens and pigs, and usually in other species
Sodium	-It acts as an extracellular component through a sodium pump that depends on energy. -Together with potassium and magnesium it intervenes in the maintenance of osmotic pressure. -Acts in acid base maintenance (ph). -Involved in the transfer of nerve impulses through the energetic potential that is associated with their separation of potassium in the cell membrane.	- Decreased appetite. - Weight reduction - Decrease in milk production. - Hemoconcentration. - Decrease in plasma volume. - Decreased urinary excretion of amino acids. - Abnormal consumption of soil, wood or sweat.
Chloride	-Control of extracellular osmotic pressure. -Maintenance of acid base balance (pH). -It is the main anion of gastric juice where it joins with hydrogen ions to form hydrochloric acid (HCl).	-Decrease in the growth rate. -In chickens after a sudden noise they fall forward with their legs extended towards the back
Sulfur	Involved in the biosynthesis of taurine, heparin and cystine. Inorganic SO4 acts on the acid base balance.	-Poor utilization of nitrogen compounds in the rumen decreasing the synthesis and / or activity of the microbial mass and, therefore,

-It intervenes in the synthesis of proteins forming amino acids. -Involved in the synthesis of lipids as a component of the vitamin biotin. -In the metabolism of carbohydrates as a component of the vitamin thiamin. -Involved in the energy metabolism as a component of the vitamin coenzyme A. -In the metabolism of collagen and connective tissue as a component of mucopolysaccharides. -In blood coagulation as a component of heparin. -Involved in the protection of peroxides from cells as a component of glutathione peroxidase.	decreasing the speed of digestion of nitrogen in the rumen. -Decrease in the production of proteins containing sulfur amino acids.

Absorption of macrominerals

Even when some Ca is absorbed in the rumen, the greatest absorption takes place in the small intestine of the deer. Absorption occurs through passive and active transport, the latter assisted by vitamin D. When the diet of the deer is relatively low in Ca, the proportion absorbed increases, on the contrary, when the diet contains more Ca than required, the absorbed proportion decreases. The absorption of P from the diet of the deer occurs mainly in the small intestine, with very little, if any, absorption in the rumen. Phosphate, like N, is recycled in the rumen through saliva and, in this way, it is incorporated into the microbes off the rumen. The recycling of P in saliva is a characteristic of ruminant animals and is a important factor in the homeostatic control of P.

Na and K are absorbed mainly in the upper part of the small intestine and, to a lesser extent, in the rumen, abomasum, the lower small intestine and the large intestine. Normally, K is absorbed from 80 to 95%. Na is absorbed in the lower parts of the intestines only if there are metabolic needs. Both elements, but especially Na, are recycled through the digestive system, mainly by secretion in saliva. The high concentrations of condensed tannins that are found in shrubs consumed by white-tailed deer in northeastern Mexico can reduce Na absorption and retention. Mg absorption occurs mainly in the rumen and omasum and to a lesser degree in the small and large intestines. The absorption of Mg is reduced by a high concentration of K, probably because the Na is essential for the transfer of Mg through the ruminal wall of the deer, and the K tends to displace the Na. Similarly, a deficiency of Na can restrict the absorption of Mg.

Parathyroid hormone (PTH)

Secreted by the gland of the same name, it is a control factor for homeostasis of Ca and P in deer. Its main sites of action are the kidneys, the gastrointestinal tract, the muscles and the bones. The most important action of the hormone is to raise the levels of Ca in the plasma. To fulfill this function, PTH acts on the skeleton, through a series of complex mechanisms that effect bone modeling and remodeling and the passage of calcium through the bone membrane. On the kidney, increasing the tubular resorption of Calcium and decreasing that of Phosphorus. On the metabolism of vitamin D, stimulating the synthesis of 1,25 (OH) 2 D3 and secondarily, the intestinal absorption of Calcium. The action on the bone: a) On the surface osteocytes and the exchange of calcium between the bone extracellular liquid (BEL) and the systemic, b) On the osteocytes of depth and c) On the remodeling and bone modeling through the osteoclasts and osteoblasts. Therefore, its action is inversely proportional to the rate of the mineral found in the plasma. The hormone acts synergistically with vitamin D by reabsorbing the Ca from the bony structures. If the Ca present in the organism is below normal levels, the action of the hormone increases and the gland hypertrophies, producing secondary nutritional hyperparathyroidism. The proper balance of calcium, sodium, potassium and magnesium ions maintains muscle tone and controls nervous irritability. the addition and removal of phosphate groups to proteins, phosphorylation and dephosphorylation, respectively, is the main mechanism for regulating the intracellular proteins and in this way the metabolism of eukaryote cells like spermatozoids.

Adult males store minerals in their skeleton and transfer them to the antlers if required, which may help explain why studies of mineral requirements do not match. In fact, while antlers are mineralizing, males suffer from osteoporosis (loss of minerals in the bones). Once the antlers harden, the minerals, such as Ca, lost from the bones are replaced by those from the diet of the deer. In any case, for a true and impressive antler, the genes of the deer play a very important role as minerals do in the diet.

Content and requirements of deer macrominerals are shown in Table 9.3. The typical concentrations in body tissues and antlers and the requirements of the main essential macro minerals of the deer are shown. The total content of the minerals in the deer without the antlers represents only 5.0%, whereas the macrominerals in the antlers compose between 30 to 35%. And, as shown in Table 93. Ca and P constitute almost 95.0% of the antler minerals. To maintain the functioning of the microbes in the rumen, the homeostasis in the organism, the growth and development of the newly weaned fawns and the growth and development of the antlers of adult males, the deer must consume a diet containing a mixture of plants with sufficient quantities of macrominerals

Table 9.3. Typical concentrations in the organism and requirements of Ca, P, Mg, K and Na of the white tail deer for growth and development of the antlers

Mineral	Body content, g/kg	Antler content, g/kg	Requirements
Calcium	15.0	190.1	4.5*
Phosphorous	10.0	101.3	2.8*
Magnesium	0.4	10.9	1.0*
Potassium	2.0	<1.0	6.0*
Sodium	1.6	5.0	1.0*

DM = dry matter
*Requirements (g/kg in the DM of the diet) to cover growth, development of the skeleton and antlers and development of the newly weaned fawns.

Macrominerals content in the plants that deer consumes

In general, the leaves of the plants contain a higher content of minerals than the stems and when the maturity progresses, the protein increases, and the minerals decrease. Apparently, the forage of native shrub plants that grow in northeastern Mexico and that are consumed by the white-tailed deer contains Ca in sufficient concentrations, throughout the year, to satisfy their requirements in all physiological states (Table 9.4). By far, the deer consuming, either individually or any mixture of these species, most likely would not suffer from any of the deficiency symptoms. Native grasses contain higher amounts of Ca than shrubs (Table 9.4) and native grasses (Table 9.4) that grow in northeastern Mexico. In addition, shrubs, forbs, cacti and flowers and fruits of northern Mexican plants contain Ca levels to meet the demands of small ruminants in any physiological state (Table 9.5).

Table 9.4. Seasonal content of Ca, P, Mg, K and Na in native plants of the northeast of Mexico that consumes the white tail deer

Mineral	Winter	Spring	Summer	Fall
		Shrubs, g/kg of dry matter		
Calcium	27.0	22.0	25.0	27.0
Phosphorous	1.0	1.2	1.2	1.0
Magnesium	6.0	6.0	6.0	5.0
Potassium	13.0	13.0	16.0	13.0
Sodium	0.5	0.4	0.5	0.4
		Forbs, g/kg of dry matter		
Calcium	29.0	28.0	37.0	31.0
Phosphorous	1.7	1.4	1.7	1.7
Magnesium	4.4	5.3	8.1	5.8
Potassium	20.0	17.0	26.0	24.0
Sodium	0.5	0.5	0.9	0.9
		Grasses, g/kg of dry matter		
Calcium	6.0	6.0	6.0	6.0
Phosphorous	1.0	1.1	1.3	1.1
Magnesium	1.4	1.3	1.7	1.6
Potassium	10.0	10.0	19.0	13.0
Sodium	0.4	0.4	0.5	0.4

All shrubs, grasses and grasses (Table 9.4) consumed by white-tailed deer in northeastern Mexico, contain P, throughout the year, in unsatisfactory amounts to meet their needs for maximum growth and development. Therefore, P is a limiting nutrient in northeastern Mexico and south Texas, USA for the optimal growth and development of newly weaned fawns, pregnant and lactating females and for maximum growth of adult male antlers. However, it has been determined that the deer does not show the deficiency symptoms characteristic of P deficiency. The above is probably due to the fact that the deer consumes herbs with a high P content, as long as they are available in the pasture. In addition, it is likely that the deer has mechanisms of conservation and transfer of P from the bone tissue to the antlers, like those known from Ca. Such mechanisms could allow the deer to select high-P herbs in the spring and preserve the P for critical periods.

Apparently. the Mg and K contained in shrub and native grasses that grow in northeastern and northern Mexico and southern Texas, USA and that are consumed by white-tailed deer (Table 9.5), are not limiting for growth and development of domestic and wild ruminants, such as white-tailed deer and that develop under extensive systems. However, native grasses (Table 9.5) have marginally seasonal (in winter) concentrations of Mg and K. However, this does not represent a pathological problem for deer because native grasses do not represent a quantitatively important component (< 1.0%) of your diet.

Table 9.5. Content of macrominerals in native plants of the state of Durango, Mexico

Plants	Calcium	Phosphorous	Sodium	Magnesium	Potassium
Shrubs	24	2	1	4	12
Forbs	21	3	1	3	16
Cacti	96	1	1	7	14
Flowers and fruits	29	3	1	4	20

Na is the most commonly deficient mineral in temperate climate ecosystems such as those in northeastern Mexico and southeastern USA and, it is the only nutrient by which, herbivores specifically develop a high appetite. Apparently, all shrubs, in all seasons of the year consumed by the white-tailed deer in northeastern Mexico and southern Texas, USA, are deficient in Na. Forbs (Table 9.4), however, only in spring and summer, when they are most abundant, contain Na in sufficient quantities to meet the needs of white-tailed deer. The same tendency is shown by native grasses (Table 9.4), but in autumn, when high precipitations promote the growth of annual pastures or the regrowth of perennials. Therefore, Na is a limiting nutrient for white-tailed deer that occurs in the northeast and north (Table 9.5) of Mexico and south Texas, USA. However, the apparent deficiencies of the deer, especially in winter and summer, can be covered with the supply of common salt (NaCl) in the salting grounds that are commonly used for domestic livestock.

Secondary compounds and mineral consumption

The temporary shortage of herbaceous vegetation in northeastern Mexico and south Texas requires that the white-tailed deer eat a diet rich in shrubs for most of the year. It is probable that deer seek minerals during periods of high consumption of shrubs so that they serve as regulators or precursors to form conjugates in the detoxification of secondary compounds contained in many of the shrub species. When feeding 4 male deer with diets consisting of 0, 25, 50 and 75% of *Acacia berlandieri*, it was found that the concentration in the diet of Ca, P, and Na decreased with the increase in the consumption of *A. berlandieri*, while the concentration of Mg did not vary. Losses of Ca, P, and Mg occurred mostly through feces, whereas losses of Na occurred via urine. The rates of consumption of Ca, Mg, and Na in diets that consisted of up to 100% of *A. berlandieri* exceeded what was required. During the summer and autumn, the adult male obtained the required P with diets consisting of 100% *A. berlandieri* and obtained the necessary during the spring and throughout the year, with diets of <75 and 97% of *A. berlandieri*, respectively. They concluded that by supplementing P during periods of low precipitation and high consumption of *A. berlandieri* can reduce the deficit of P in females in the reproductive period.

The condensed tannins can chelate strongly mineral elements, which reduces their absorption and increases the endogenous losses of the digestive tract. Plants that contain either condensed or hydrolyzable tannins can lower the body content of Na. The supplementation with minerals in diets with tannins prevented the drastic reduction of Na and decreased the toxic effects caused by the consumption of tannins. Similar effects were obtained when food containing tannins and saponins was provided. Apparently, a triterpene saponin did not cause the drastic decrease in Na. It is known that Na is limiting for some herbivores that frequently exhibit a strong need to consume Na. The limitations for the consumption of Na are closely related to the consumption of tannins and other secondary compounds that probably cause the drastic reduction of Na. The supplementation of Na or other minerals provides a mechanism by which plants maintain the carrying capacity of herbivores below the levels where they cannot cause severe damage to vegetation.

Microminerals

The difference between the trace elements (microminerals) and the macrominerals is based on the relative amounts that the deer needs from each one in the diet for a normal operation. It has been reported that Co, Cu, I, Fe, Mn, Mo, Se and Zn are essential for the functioning of the normal metabolism of ruminants such as deer. In general, they act as activators of the enzyme systems or as components of the organic compounds and, as such, are needed in small quantities (ppm = mg/kg). The functions and symptoms of deficiency of the trace elements are shown in Table 9.6.

Absorption of microminerals

The trace elements are absorbed mainly in the small intestine and, sometimes, to a lesser degree, in the large intestine. In general, the degree of absorption depends on the balance between the supply in the diet and the metabolic demand. Absorption in excess is prevented by homeostatic mechanisms. The homeostatic control of the absorption is particularly important in the Fe, since the deer have limited capacity of excretion. The absorption is also influenced by the physiological status of the deer, by the chemical form of the ingested element and by other components of its diet. During growth, lactation and pregnancy, there is an increase in demand, particularly for Fe, Mn and Zn, and therefore, an increase in the percentage of absorption. In relation to other components of the diet, a high consumption of any of the micronutrients cations is likely to interfere with the absorption of others, although not necessarily at all. For example, Mn, Zn and Cu interfere in the absorption of Fe, presumably by competition for binding sites in the digestive tract, and a high Fe consumption depresses the absorption of Cu.

Content and requirements of microminerals

Tissue content and Cu, Mn, Fe and Zn requirements in ruminants, such as white-tailed deer, are shown in Table 9.7 It should be noted that the requirements of trace elements of white-tailed deer have not yet been determined; however, for comparison reasons, the average requirements of cattle, sheep and goats will be used. In addition, the requirements of trace elements of the deer, depend on factors such as sex, age, activity and the environment where it develops. However, the recommended concentrations include an appreciable margin of safety for the deer not to suffer symptoms of deficiency in their different physiological states including the growth and development of their antlers.

Table 9.6. Functions and symptoms of deficiency of essential trace elements by the deer organism

Mineral	Functions	Symptoms de deficiency
Cobalt	- The only known function is as a component of vitamin B12 (cyanocobalamin)	• Acts as an enzyme in several enzyme systems, among which isomerases and dehydrogenases stand out. • Involved in the biosynthesis of methionine. • Involved in the oxidation of propionic acid. • It is needed as a component of the coenzymes that are required for the synthesis of methyl groups and their metabolism. • Together with folic acid, they act in the synthesis of nucleoproteins. • Involved in the metabolism of the amino acid leucine
Copper	• It is needed for the activity of enzymes that are associated with iron metabolism. • It is required for the production of elastin. • Needed for the production of collagen. • They are required for the production of melanin. • It is used for the integrity of the central nervous system. • Prevents the oxidation of cells. • Involved in the synthesis of the atp	-The gradual decrease of copper produces anemia due to the association with iron. • Loss of hair pigmentation. • Increase in susceptibility to infectious diseases. It has been found that a high consumption of molybdenum and sulfur can form together with copper an insoluble salt that causes a poor absorption of copper
Iodine	• It is part of the proteins that contain iodine and are found in the thyroid, including mainly the thyroglobulins that produce the hormones triiodothyronine (T3) and tetraiodothyronine (T4)	• Decrease in the basal metabolic rate, due to the control of the oxidation index; In young animals the disease is called cretinism and in adult myxedema. • Goiter consisting of an increase in the size of the cells of the thyroid gland.
Iron	• In the transport of electrons as a component of a large number of enzymatic systems, among which metalloporphyrins and metaloflavinics. • In oxygen transport and storage component of two important proteins such as hemoglobin and myoglobin	• The most common symptoms in animal organisms is anemia, of the hypochromic microcytic type which means small and few red cells.
Manganese	• It is essential for the formation of chondroitin sulfate, which is a component of the mucopolysaccharides of the organic matrix of bone. • Avoid ataxia in animals. • It is a component of several enzymes that act in the metabolism of polysaccharides, glycoproteins, carbohydrates and lipids.	• Skeletal abnormalities (lameness, shortening and arching of the legs, and enlargement of the joints) associated with the lack of Mn in the mucopolysaccharides.
Molybdenum	• It is the component of the enzymes xanthine oxidase, sulfite oxidase and aldehyde oxidase	• Low levels of xanthine oxidase. • No deficiency symptoms have been reported.
Selenium	• It is a component of the enzyme glutathione peroxidase, which is involved in the catabolism of peroxides that originate in the oxidation of tissue lipids and, therefore, has a central role in the integrity of cell membranes. • It is a component of the enzyme iodothyronine deiodinase type 1.	• Nutritional muscular dystrophy in ruminants

Zinc	• It is a constituent of numerous metalloenzymes including carbonic anhydrase, carboxypeptidases a and b, various dehydrogenases, alkaline phosphatase, ribonuclease and DNA polymerase. • It is needed for the normal synthesis of proteins and for their metabolism. • It is a component of insulin and thus acts on the metabolism of carbohydrates. • Acts in gene expression. • Intervenes in the stability of the membrane.	• Anorexia in all species. • Thickening or hyperkeratinization of epithelial cells. Delayed bone formation. • Hypogonadism is observed in males

Content of microminerals in the plants that the deer consumes

Most of the Cu in the deer organism is located stored in the liver (around 80%). However, white-tailed deer require 8 mg/kg of Cu in the dry matter of their daily diet (Table 9.7). Although the amount required is too small to perform the metabolic functions that require Cu, some shrub species, which are consumed by the white-tailed deer that occurs in northeastern Mexico and southern Texas (Table 9.8), have concentrations marginally lower than their Cu requirements,

In contrast to the fact that a significant number of bush forages contain marginally low levels of Cu, all of the grasses consumed by white-tailed deer (Table 9.8) contain Cu in concentrations sufficient to meet the requirements of white-tailed deer. Apparently, all the native grasses (Table 9.8) that are consumed by the white-tailed deer, have Cu concentrations, in all the stations with insufficient levels to satisfy the requirements of the white-tailed deer.

The Mn is uniformly distributed throughout the deer organism, although with some enrichment in the liver, bone tissue and the digestive system. The requirements of deer Mn are 30 mg/kg in the dry matter of their diet. With the exception of native shrub species such as *Acacia rigidula, Cercidium macrum, Acacia farnesiana, Porlieria angustifolia, Celtis pallida, Acacia berlandieri, Leucaena leucocephala, Leucophyllum texanum, Acacia greggii, Cordia boissieri, Condalia obovata, Prosopis glandulosa*, and *Opuntia engelmannii*, other 19 plants (Table 9.8) have concentrations of Mn to satisfy the metabolic requirements of Mn of white-tailed deer that grows in northeastern

Table 9.7. Typical concentrations in the organism and Cu, Mn, Fe and Zn requirements of white-tailed deer for its growth and development

Mineral	Body content, mg/kg	Requirements, mg/kg
Copper	1.0 – 5.0	8*
Manganese	0.2 – 0.5	30*
Iron	20.0 – 80.0	40*
Zinc	10.0 – 50.0	30*

DM = Dry matter * Requirements (mg/kg in the DM of the diet) to cover growth, development of adult ruminants such as white-tailed deer.

Table 9.8. Seasonal content of Cu, Fe, Mn, and Zn in native plants of northeastern Mexico consumed by white-tailed deer

Mineral	Winter	Spring	Summer	Fall
	\multicolumn{4}{c}{Shrubs, g/kg in dry matter}			
Copper	9	9	10	9
Iron	157	143	160	159
Manganese	84	76	82	71
Zinc	37	38	43	40
	Forbs, g/kg in dry matter			
Copper	12	11	16	15
Iron	189	181	421	275
Manganese	43	51	62	57
Zinc	38	36	65	50
	Grasses, g/kg in dry matter			
Copper	3	4	5	4
Iron	112	132	185	163
Manganese	37	31	44	41
Zinc	40	41	55	45

Apparently, the native forbs that are selected by the white-tailed deer in northeastern Mexico and shown in 9.8, Have sufficient concentrations of Mn, during all seasons of the year, to cover the requirements of the white-tailed deer. In addition, the native grasses shown in Table 9.8 and, which also make up the group of plants that select white-tailed deer in northeastern Mexico, have concentrations of Mn that exceed the metabolic demands of white-tailed deer. *Hilaria belangeri* is a perennial grass native to the ecosystem flora of the Tamaulipas Scrub that includes the states of Coahuila, Nuevo León and Tamaulipas and the south of Texas, USA and that is consumed by the Texas white-tailed deer, besides its foliage is a good source of microminerals including Mn

Iron is a mineral found in abundance in all types of plants selected by white-tailed deer in northeastern Mexico and southern Texas, USA. All shrubs and those shown in Table 9.8 have Fe concentrations that cover and in many of them exceed the Fe's requirements of white-tailed deer. Also, forbs and grasses (Table 9.8) contain Fe levels that meet the requirements of the deer.

According to studies of selectivity of the deer, its diet consists of a mixture of shrubs, forbs and grasses, with a tendency to a predominance of shrubs. In addition, the occurrence of forbs in the pasture occurs when there is sufficient humidity and, usually, only occurs in summer and autumn. Therefore, it is not very likely that the deer show symptoms of toxicity caused by an excess in the consumption of forbs with high concentrations of Fe. However, high concentrations of condensed tannins in the plants that the deer consumes can diminish the absorption of Fe.

Table 9.9. Content of microminerals in native plants of the state of Durango, Mexico

Plants	Cu	Fe	Mn	Zn
Shrubs	8	197	52	31
Forbs	9	257	60	45
Cacti	8	82	154	40
Flowers and fruits	7	108	270	44

In summer alone, shrubs, grasses and grasses consumed by white-tailed deer in northeastern Mexico (Table 9.8) contain Zn concentrations to satisfy deer requirements. In other seasons the levels are marginally satisfactory (30 mg/kg). In addition, herbs, which the deer consumes in northern Mexico, exceed in their Zn content (Table 9.9).

Chapter 10

Vitamins

> **Abstract.-** The vitamins according to their properties are divided into water soluble and fat soluble. They are a group of complex organic compounds present in tiny amounts in natural foods that are essential for the normal metabolism of animals and, the lack of them in the diet, causes a symptom of deficiency. The fat solubles can be stored in the organism of the deer and are excreted mainly in feces via bile, are composed mainly of carbon, hydrogen and oxygen. The fat-soluble vitamins A, D and E of vitamins are the most important for the proper growth and development of the deer and K intervenes in blood coagulation. Water soluble vitamins, with the exception of cobalamin (vitamin B12), are not stored in the tissues of the deer in appreciable amounts; however, the microbial flora of the deer rumen synthesizes, to a large extent, the required amounts of water-soluble vitamins and vitamin K. Most act as metabolic catalysts, usually as coenzymes. All cause severe failures in the metabolism of deer if they are not available in sufficient quantities in their tissues. Most water soluble vitamins are unrelated, they are required in very small amounts, with the exception of choline which is part of the phospholipid molecule. The main antioxidant vitamins such as vitamin C, vitamin E and α-carotene (vitamin A) serve to stabilize these highly reactive free radicals, and in this way maintain the functional and structural integrity of the cells. Therefore, antioxidants are very important in the defense of the immune system and the health of the deer.

Introduction

The vitamins according to their properties are divided into two large groups: 1) water-soluble vitamins (soluble in water) and 2) fat-soluble vitamins (soluble in lipids). Vitamins are defined as a group of complex organic compounds present in minute amounts in natural foods that are essential for the normal metabolism of animals and, the lack of them in the diet, causes a symptom of deficiency. However, some vitamins do not necessarily fall within this definition because some do not precisely come from food. Certain substances that are considered as vitamins are synthesized by the bacteria of the intestinal tract in quantities that may be adequate for the needs of the organism. However, there is a clear distinction between vitamins and substances that are synthesized in the tissues of the body. Ascorbic acid, for example, is produced by most animal species including deer, except in young animals or when under stress. Also, the water-soluble vitamin niacin can be synthesized from the amino acid tryptophan and vitamin D that is synthesized from a precursor that is found in the skin and is activated by the action of ultraviolet light. Therefore, under certain circumstances and for certain species, vitamin C, niacin and vitamin D do not always conform to the classical classification of vitamins.

Fat-soluble vitamins

Lipid soluble vitamins can be stored in the deer organism. These are excreted mainly in feces via bile, are composed mainly of carbon, hydrogen and oxygen. They are absorbed by passive diffusion through the lipid phase of the mucosal cell membrane. The functions and symptoms of deficiency are shown in Table 10.1. Vitamins A, D and E are among the vitamins most important for the proper growth and development of the deer.

Vitamin A

It is not found as such in vegetables, but as its precursor, carotene. It is commonly known as provitamin A because the body can convert it into its active form. Retinol is alcohol, retinal is aldehyde, and retinoic acid is the acid of vitamin A. Vitamin A is undoubtedly very important for the growth and hardness (ossification) of deer antlers. In addition, the deer can convert the carotenes into vitamin A. During most of the year, the consumption of carotenes is sufficient, but small deficiencies of Vitamin A can occur during times of drought or prolonged winters. Vitamin D is important to promote the absorption of Ca, especially during the growth and hardening of the antlers.

Fats facilitate the absorption of both vitamin A and carotene, emulsifying agents have additional effects, some of the provitamins that are ingested are destroyed in the intestine, vitamin E, an antioxidant, in food, decreases this destruction; Vitamin A in the diet comes in the form of long-chain retinyl esters. These esters are hydrolyzed in the intestinal rumen with the pancreatic retinyl ester hydrolase or by a hydrolase in the intestinal mucosa. In some species such as: rat, pig, goat, sheep, rabbit, zebra, donkey, bison and dog, almost all the carotene is transformed in the intestine; In man, cattle, horses and carp can absorb significant amounts of carotene. This can be stored in liver and fatty tissues. Therefore, these animals have body fat and yellow milk, animals that do not absorb carotene have white fat, in some species the carotene can be converted into vitamin A by various body tissues other than the intestine; in ruminants, the conversion efficiency of absorbed carotenoids to vitamin A is very low. In the cells of the liver parenchyma, a large amount of vitamin A is stored, such as retinyl ester and mainly as palmitate. The liver can store enough vitamin A to protect the animal during prolonged periods of scarcity.

Vitamin D

Vitamin D or calciferol is an unsaponifiable heterolipide of the group of steroids. It is also called antirachitic vitamin because its deficit causes rickets. It is a provitamin soluble in fats and can be transformed by cholesterol or ergosterol (own of vegetables) by solar radiation. It is responsible for regulating the passage of calcium (Ca 2+) to the bones. Therefore, if vitamin D is missing, this step does not occur, and the bones begin to weaken and curl causing irreversible malformations, rickets. In addition, it plays an important role in the maintenance of organs and systems through multiple functions, such as: the regulation of calcium and phosphorus levels in the blood, promoting the intestinal absorption of them from food and the reabsorption of calcium at the renal level. With this it contributes to bone formation and mineralization, being essential for the development of the skeleton. However, in very high doses, it can lead to bone resorption. It also inhibits the secretions of parathyroid hormone from the parathyroid gland and affects the immune system due to its immunosuppressive role, promotion of phagocytosis and antitumor activity. Vitamin D deficiency can result from the consumption of an unbalanced diet, coupled with inadequate sun exposure; it can also occur due to disorders that limit its absorption, or conditions that limit the conversion of Vitamin D into active metabolites, such as alterations in liver or kidney, or rarely in some hereditary disorders. Vitamin D deficiency causes a decrease in bone mineralization, leading to soft bone diseases, such as rickets in children and osteomalacia in adults, and is even associated with the appearance of osteoporosis. The requirements of vitamin D or calciferol are covered, most likely by the exposure of the deer to sunlight (ultra violet rays) and by the consumption of tissue from plants irradiated with ultra violet light. Vitamin D is a prohormone, so it has no hormonal activity by itself, but it is converted to active hormone 1,25-D through a highly regulated synthesis mechanism.

Vitamin E

The alpha-tocopherol or vitamin E acts as an antioxidant at the level of the synthesis of the heme pigment, which is an essential part of the hemoglobin of the red blood cells. Vitamin E in its natural state has about eight different forms of isomers, four tocopherols and four tocotrienols. All isomers have aromatic rings with a hydroxyl group which can donate a hydrogen atom to reduce the free radicals of the materials that make up the hydrophobic biological membranes of the cell walls. There are alpha α, beta β, gamma γ and delta δ forms for both isomers, and it is determined by the number of methyl groups in the aromatic ring. Each of the forms has its own biological activity. Vitamin E is important to prevent muscle tissue damage when the deer is subjected to hardly efforts.

Vitamin K

Vitamin K is important for the blood coagulation of deer. Its deficiency is not likely to occur because the green leafy plants that the deer consumes contain it, and the microflora of the rumen can synthesize it in sufficient quantities so that the deer does not manifest hemorrhagic syndrome. All members of the Vitamin K group share a methylated ring of naphthoquinone in its structure, and it varies in the aliphatic side chain attached in the 3rd position. Phylloquinone (also known as Vitamin K1) invariably contains in its side chain 4 isoprenoid residues, one of which is unsaturated. Menaquinones have a side chain composed of a variable number of unsaturated isoprenoid residues, generally designated MK-n, where n specifies the number of isoprenoids. It is generally accepted that naphthoquinone is the functional group, so the mechanism of action is similar for all forms of Vitamin K. Substantial differences may be expected, however, with respect to intestinal absorption, transport, distribution to tissues and bioavailability. These differences are caused by the different lipid affinities of the side chains, and by the various matrices of the food in which they occur.

Table 10.1. Functions and symptoms of deficiency of fat-soluble vitamins in the deer organism

Vitamins	Functions	Symptoms of deficiency
Vitamin A1 Retinol, Retinal, Retinoic acid). Vitamin A2 (Dehydroretinol)	• It is required to have normal night vision. • In the formation of the protein-pigment rhodopsin of the visual purple of the eye. -It is needed in the normal epithelial cells that cover or cover the surfaces or body cavities such as the digestive, urinary and respiratory tracts and those of the skin. • It has an important function in the synthesis of glycoproteins.	• Night blindness. • Xerophthalmia in growing animals. the disease is characterized by irritation of the cornea and the conjunctiva and produces cloudiness and infection
Vitamin D2 (Ergocalciferol) Vitamin D3 (Colecalciferol)	• Along with parathyroid hormone, promotes the absorption of Ca in the digestive system. • Promotes the synthesis of the calcium-carrying enzyme.	-Tetanus hypocalcemia. • Abnormal skeletal development. • Rickets in young animals. • Osteomalacia in adult animals.
Vitamin E (Tocopherol, Tocotrienols)	• It is an antioxidant (oxido-reducer) or biological predator of free radicals (OH) in the metabolism of nucleic acids and proteins and mitochondrial metabolism. • Helps maintain the compartmentalization and permeability of the membranes. • Controls the synthesis of prostaglandins in some tissues. • Intervened in the synthesis of some specific proteins.	-Failures in reproduction. -Disruption of cell permeability. -Muscular injuries (myopathies). The disease is called exudative diathesis.
Vitamin K1 (Phylloquinone) Vitamin K2 (Menaquinone) Vitamin K3 (Menadiona)	• It is needed for the synthesis of prothrombin in the liver and together with the factors II, VII, IX and X, it is necessary to carry out normal blood coagulation.	• Prolongation of blood coagulation time, generalized hemorrhages and death in severe cases. • Mycotoxins (fungal toxins) can be vitamin k antagonists.

Water-soluble vitamins

Water-soluble vitamins, except for cobalamin (vitamin B12), are not stored in the tissues of the deer in appreciable amounts; however, the microorganisms of the deer rumen synthesize, to a large extent, the required amounts of water-soluble vitamins and vitamin K. Most act as metabolic catalysts, usually as coenzymes. All cause severe failures in the metabolism of deer if they are not available in sufficient quantities in their tissues. Most water-soluble vitamins are unrelated, they are required in very small quantities, except for choline which is part of the phospholipid molecule (lecithins). The functions and symptoms of deficiency are shown in Table 10.2.

Thiamin

Vitamin B1, also known as thiamin, is a molecule consisting of two interconnected organic cyclic structures: a pyrimidine ring with an amino group and a sulfurized thiazole ring attached to the pyrimidine by a methylene bridge. It is soluble in water and insoluble in alcohol. Its absorption occurs in the small intestine (jejunum, ileum) as free thiamine and as thiamine diphosphate (TDP), which is favored by the presence of vitamin C and folic acid but inhibited by the presence of ethanol (alcohol). It is necessary in the daily diet of most vertebrates such as the deer and some microorganisms such as those of the rumen. Its lack in man causes a disease known as beriberi (Table 10.2).

Riboflavin

Vitamin B2, so called in the first instance, undoubtedly contained a mixture of developmental factors, one of which was isolated and turned out to be a yellow pigment that is now known as riboflavin. Riboflavin is still sometimes referred to as vitamin B2, which is not strictly correct. Riboflavin belongs to the group of fluorescent yellow pigments called flavins. It is an easily absorbed micronutrient, with a key role in the maintenance of animal health. It is the main component of the cofactors FAD and FMN and is therefore required by all flavoproteins, as well as for a wide variety of cellular processes. Like other B vitamins, it plays an important role in energy metabolism, and is required in the metabolism of fats, carbohydrates and proteins (Table 10.2). Vitamin B2 is a yellow water-soluble vitamin, consisting of a dimethylated isoaloxazine ring to which ribitol, an alcohol derived from ribose, binds. The three rings form the isoaloxacin and the ribitol is the chain of 5 carbons in the upper part. This vitamin is sensitive to sunlight and to certain treatments, such as pasteurization, a process that causes 20% of its content to be lost. For example, exposure to sunlight from a glass of milk for two hours causes 50% of the vitamin B2 content to be lost.

Niacin

Vitamin B3, niacin, nicotinic acid or vitamin PP is a water-soluble vitamin whose derivatives, NADH and NAD +, and NADPH and NADP +, play essential roles in the energy metabolism of cells and DNA repair. The designation vitamin B3 also includes the corresponding amide, nicotinamide, or niacinamide, with chemical formula $C_6H_6N_2O$. The functions of niacin include the removal of toxic chemicals from the body and the participation in the production of steroid hormones synthesized by the adrenal gland (Table 10.2), such as sex hormones and hormones related to stress.

Vitamin B6

Vitamin B6 is actually a group of three chemical compounds called pyridoxine (or pyridoxol), pyridoxal, and pyridoxamine. The phosphorylated derivatives of pyridoxal and pyridoxine (pyridoxal phosphate (PLP) and pyridoxamine phosphate (PMP) respectively) have coenzyme functions. They participate in many enzymatic reactions of amino acid metabolism and their main function is the transfer of amino groups (Table 10.2); therefore, they are coenzymes of transaminases, enzymes that catalyze the transfer of amino groups between amino acids that coenzymes act as temporary transporters of amino groups.

Pantothenic acid

Vitamin B5 or pantothenic acid is a water-soluble vitamin required to maintain life (essential nutrient). Pantothenic acid is needed to form the coenzyme A (CoA) and is considered critical in the metabolism and synthesis of carbohydrates, proteins and fats (Table 10.2). In its chemical structure it is an amide between D-pantothenate and beta-alanine. Its name derives from the Greek Pantothen, which means "from all over", and small amounts of pantothenic acid are found in almost all foods, with high amounts in whole grain cereals, legumes. It is commonly found as an alcohol analogue, provitamin panthenol and calcium pantothenate.

Biotin

It is a vitamin stable to heat, soluble in water, alcohol and susceptible to oxidation that intervenes in the metabolism of carbohydrates, fats, amino acids and purines (Table 10.2). Biotin is found in the cell bound with specific residue of lysine (an amino acid) forming the biocytin; Biocytin is covalently linked to certain enzymes related to the formation or use of carbon dioxide, and thus exerts a coenzyme function: it acts in the transfer (acceptor and donor) of carbon dioxide in numerous carboxylases:

- Acetyl-CoA carboxylase alpha and beta
- Methylcrotonyl-CoA carboxylase
- Propionyl-CoA carboxylase
- Pyruvate carboxylase

All these enzymes are essential in the processes of cellular duplication. Biotin is used in cell growth, the production of fatty acids and in the metabolism of fats and amino acids. Play a role in the citric acid cycle, which is a process by which biochemical energy is generated during aerobic respiration. Biotin not only assists in several chemical and metabolic conversions, but also helps transfer carbon dioxide. Biotin also participates in the maintenance of blood glucose levels.

Folic acid

Folic acid does not have coenzyme activity, but its reduced form, tetrahydrofolic acid, frequently represented as FH4. It acts as an intermediate transporter of groups with a carbon atom, especially formyl groups, which is required in the synthesis of purines, compounds that are part of the nucleotides, substances present in DNA and RNA, and necessary for their synthesis during the phase S of the cell cycle, and therefore for cell division; it also acts in the transfer of methenyl and methylene groups. Folate is necessary for the production and maintenance of new cells (Table 10.2). This is especially important during periods of rapid cell division and growth as in childhood and pregnancy. Folate is necessary for DNA replication. Because of this, folate deficiency hinders cell synthesis and division, mainly affecting the bone marrow, a site of rapid cell turnover. Because the synthesis of RNA and proteins is not completely impeded, long or non-regular blood cells called megaloblasts form, resulting in megaloblastic anemia.

In the form of a series of tetrahydrofolate components, folate is derived as a substrate in a number of reactions and is also involved in the synthesis of dTMP (2'-deoxythymidine-5-phosphate) from dUMP (2'-deoxyuridine-5- phosphate). It helps convert vitamin B12 into one of its coenzyme forms and participates in the DNA synthesis required for rapid cell growth. The pathways leading to the formation of tetrahydrofolate (FH4) begins when the folate (F) is reduced to dihydrofolate (FH2), which is then reduced to tetrahydrofolate (FH4). Dihydrofolate reductase catalyzes both steps.

Methylene tetrahydrofolate (CH2FH4) is formed from tetrahydrofolate with the addition of methylene groups from one of the donor carbons: formaldehyde, serine or glycine. Methyl tetrahydrofolate (CH3FH4) can be formed from methylene tetrahydrofolate by reduction of the methylene group; the formal tetrahydrofolate (CH3-FH4) results from the oxidation of methylene tetrahydrofolate.

Cyanocobalamin or vitamin B12

Vitamin B12 is a hexacoordinated cobalt complex. Four coordination positions are occupied by a macro corrina cycle. One of the axial positions is completed with a cyanide group (CN-). A side chain of the corrin ring composed of an amide, a phosphate group, a ribose and a nucleotide complete the coordination through the 5,6-dimethylbenzimidazole moiety at its end.

The coenzyme form of cyanocobalamin is deoxyadenosylcobalamin (usually called coenzyme B12), in which the cyanide group is replaced by 5'-deoxyadenosine (a nucleoside). It acts as a transient transporter of alkyl and alkyl-substituted groups, in two internal rearrangements of the same molecule, such as the transfer of groups from one carbon atom to another adjacent to the same substrate; for example, in the transformation of methylmalonyl-CoA into succinyl-CoA. Transfer of methyl groups between two different molecules, for example in the transformation of the amino acid homocysteine into methionine.

Vitamin B12 binds to a protein in the saliva called protein R, an R + B12 complex is formed until it reaches the lumen of the stomach. The parietal cells of the fundic glands of the stomach synthesize hydrochloric acid and intrinsic factor (IF). This intrinsic factor is a glycoprotein that is secreted in the cells of the stomach walls in response to the presence of histamine, gastrin and pentagastrin, which are normally found in food. In the duodenum, there are enzymes that favor the breakdown of the R + B12 complex and the binding of vitamin B12 to the IF. Vitamin B12 or cyanocobalamin is absorbed by endocytosis in cells of the terminal ileum, where enterocytes have receptors for intrinsic factor. Absorption of B12 can be active, mediated by the IF or passive independent of the IF.

Once absorbed and inside the blood vessels, it travels together with plasma proteins called transcobalamines II to reach the cells of the bone marrow and liver cells where it is stored. To enter the cells, the entire transcobalamin II-B12 complex enters, after which this binding is broken by lysozymes and these vitamins are already fully usable by the cell. Liver reserves are approximately 2-3 mg. The functions of vitamin B12 are related to the synthesis of methionine and thymidine in DNA duplication and to the synthesis of acetyl CoA for myelination of the CNS. If there is a deficit of B12 or IF, DNA synthesis will be affected, by defect in the production of purines and pyrimidines, and therefore cell duplication, can cause some kind of megaloblastic anemia.

Choline

Choline is an essential nutrient for cardiovascular and cerebral functioning, and for the cell membrane and its normal functioning (Table 11.2). It is part of acetylcholine (a neurotransmitter) and of phosphatidylcholine (a phospholipid that forms part of the plasma membrane of all cells).

Choline and its metabolites are necessary for three main physiological functions:

1. Maintenance of the structure and of certain communication mechanisms of the cell membrane.
2. Cholinergic neurotransmission (synthesis of acetylcholine)
3. Donor of methyl groups via their metabolites, trimethylglycine (betaine) that participates in the synthesis of S-adenosylmethionine

Vitamin C or Ascorbic Acid

The L (levorotatory) enantiomer of ascorbic acid is also known as vitamin C. Ascorbic acid and its sodium, potassium and calcium salts are widely used as antioxidants and additives (Table 11.2). These compounds are soluble in water, so they do not protect fats from oxidation. For this purpose, fat-soluble ascorbic acid esters can be used with long-chain fatty acids (palmitate and ascorbyl stearate).

A large majority of animals and plants can synthesize vitamin C, through a sequence of 4 enzymatic steps, which convert glucose into vitamin C. The glucose necessary to produce ascorbate in the liver (in mammals) is extracted from glycogen, therefore the synthesis of ascorbate is a glycoleptically-dependent process. In reptiles and birds, biosynthesis is carried out in the kidneys. Human beings do not possess the enzymatic capacity to produce vitamin C. The cause of this phenomenon is that the last enzyme of the synthesis process, L-gulonolactone oxidase is absent because the gene for this enzyme is defective. The mutation is not lethal to the organism, because vitamin C is abundant in food sources. It has been detected that the species with this mutation (including humans) have adapted a recycling mechanism to compensate it.

Vitamin C can be absorbed as ascorbic acid and as dehydroascorbic acid at the level of buccal, stomach and jejunal mucosa (small intestine), then transported via the portal vein to the liver and then transported to the tissues that require it. It is excreted through the kidneys (in the urine), mainly in the form of oxalic acid; only the non-absorbed vitamin is eliminated by feces. It has been observed that the loss of the ability to synthesize ascorbate is surprisingly parallel to the evolutionary loss of the ability to decrease uric acid. Uric acid and ascorbate are strong reducing agents. This has led to the suggestion that in higher primates, uric acid has assumed some functions of ascorbate.

Vitamin requirements of the deer

As previously mentioned, the plants (leaves, seeds and fruits, etc.) that the deer typically consumes and especially in the spring, summer and autumn seasons provide the amounts of vitamins necessary for the growth and development of the deer. However, during periods of dry or prolonged dry winters, the plants do not contain foliage with adequate nutritional quality to provide the required nutrients for the deer, especially for the newly weaned fawns. Therefore, for a good nutritional management of the fawns to reduce the mortality rate, micronutrient supplementation programs such as vitamins should be carried out. 5000 IU, 400 IU and 200 IU per day of vitamin A, vitamin D and vitamin E are recommended, respectively. However, fawns born to well-fed mothers have very low vitamin stores. Also, they do not have an immune system that protects them from viral and bacterial diseases. Unlike adult deer, young fawns do not have a fully functioning rumen and active ruminal microbiota, which typically contributes to the synthesis of vitamins. Colostrum, which is produced by the mammary gland during the first 5 to 10 days after the fawn is born, is rich in vitamins, particularly vitamin A, provides sufficient vitamins for the metabolic needs of the newborn fawn. Therefore, the daily supply of vitamins for newborn fawns is provided by colostrum. The latter also provides the globulins that will be integrated into the lymph to build the immune system of newly weaned fawns.

Functions of vitamins

The classic deficiency symptoms and non-specific parameters (e.g., low production and reproduction rates in deer) are associated with deficiencies or excesses of vitamins. The nutrition of the vitamins in the deer should not only be considered important to prevent deficiency symptoms, but also, to optimize the healthy productivity of the deer. Due to the importance of the above, emphasis will be placed on the role of vitamins as antioxidants, strengthening the immune system and as promoters of the productive quality of deer.

Free radicals can severely damage the biological systems of animals such as deer. Free radicals, which include the hydroxide, hypochlorite, peroxide, superoxide, hydrogen peroxide and oxygen groups, are generated by autoxidation, radiation or by activities of some oxidases, dehydrogenases and peroxidases. These oxidative products can, in turn, damage healthy cells if they are not eliminated. The main antioxidant vitamins such as vitamin C, vitamin E and α-carotene (vitamin A) serve to stabilize these highly reactive free radicals, and in this way maintain the functional and structural integrity of the cells. Therefore, antioxidants are very important in the defense of the immune system and the health of the deer. Also, several metalloenzymes such as glutathione peroxidase (selenium), catalase (iron) and superoxide dismutase (copper, zinc and manganese) also play an important role in protecting the intracellular constituents of oxidative damage. Therefore, the balance in the diet of these nutrients and in the tissues of the deer are important to protect the tissues against the damage of free radicals.

Table 10.2. Functions and symptoms of deficiency of water-soluble vitamins in the deer organism

Vitamins	Functions	Symptoms of deficiency
Thiamin (Vitamin B_1)	Thiamine is phosphorylated in the liver to form thiamine pyrophosphate (PFT) and lipothiamine (LPFT), thus it is related to the oxidative decarboxylation of alpha ketoacids (pyruvate and alpha ketoglutarate).	PFP deficiency raises the blood concentration of pyruvate and lactate. Decreased appetite
Riboflavin (Vitamin B2)	Acts in coenzymes FAD (flavin adenine dinucleotide) and FMN (flavin mononucleotide), which are present in a large number of enzyme systems.	Decrease in the growth rate in young animals. Infiltration of fat in the liver and opacity in the cornea.
Niacin (Vitamin B3)	It is a constituent of two important coenzymes that act as codehidrogenases: NAD (nicotin adenine dinucleotide) and NADP (nicotine adenine dinucleotide phosphate), and its function is to transfer hydrogen from substrates to molecular oxygen to form water (phosphorylation oxidative of ATP).	Decreased growth and appetite
Vitamin B6 (Pyridoxol,	ts as an enzyme of a large number of enzymatic systems that are associated with the metabolism of	n all species the deficiency causes seizures due to demyelination of the

Pyridoxine, Pyridoxamine)	proteins and nitrogen. The main classes of enzymatic reactions are the following: transamination, decarboxylation, among others	myelin sheaths and later degenerative changes..
Pantothenic acid (Vitamin B5)	It acts as a component of coenzyme A that is needed for the acetylation of numerous compounds that are part of the energy metabolism. Coenzyme A is needed for the formation of fragments of two carbons from fats, amino acids and carbohydrates to enter the Krebs cycle and for the synthesis of steroids.	Infiltration of fat in the liver. Rough coat of hair. Dermatitis. Anorexy. Loss of hair around the eyes. Demyelination of the sciatic nerve and the spinal cord.
Biotin (Vitamin H)	It is a component of several enzyme systems, participating in the following reactions: conversion of propionyl coenzyme A into methylmalonyl coenzyme A in the metabolism of propionic acid to become glucose in ruminants. Involved in the degradation of the amino acid leucine It participates in the metabolism of fats in transcarboxylation reactions.	Progressive desquamative dermatitis and alopecia. In males it delays the genital development.
Folic acid (Vitamin M)	Involved in a series of metabolic enzymatic reactions that involve the incorporation of individual carbon units to larger molecules. It intervenes in the synthesis of purines, pyrimidines, the amino acids glycine and serine, and creatinine. Involved in the synthesis of the enzymes xanthine oxidase and choline oxidase	Macrocytic, hyperchromic anemia with leukopenia and thrombocytopenia. Decrease in the growth rate. Decreases deer resistance to infections.
Vitamin B12 (Cianocobalamin)	Acts as an enzyme in several enzymatic systems, among which isomerases and dehydrogenases stand out. It intervenes in the biosynthesis of methionine. Involved in the oxidation of propionic acid. It is needed as a component of the coenzymes that are required for the synthesis of methyl groups and their metabolism. Together with folic acid acts in the synthesis of nucleoproteins. Involved in the metabolism of the amino acid leucine.	The absorption of vitamin B12 is conditioned by the presence of an enzyme secreted in the stomach called intrinsic factor. the absence of intrinsic factor causes deficiency. In ruminants the absence of cobalt causes B12 deficiency because it is the main component of the vitamin. In the species studied, it also causes an increase in the content of pantothenic acid and growth disorders.
Choline (Gosipina)	It is a structural component of tissues forming part of lecithins and sphingomyelins. It is related to the transmission of nerve impulses as a component of acetylcholine It supplies biologically labile methyl groups. It has lipotropic effects (movements of lipids or fats) so it prevents the accumulation of liver fat	Choline deficiency is associated with fatty liver, since choline is a component of phospholipids (lecithins).
Vitamin C (Ascorbic acid)	It is directly related to a large number of enzymes that catalyze the oxidation reactions and reduction in electron transport. In this way it acts as a powerful reducing agent. It is needed to maintain the normal oxidation of the amino acid Tyrosine.	t is not likely that the deer present symptoms of vitamin deficiency because their tissues synthesize the necessary amounts, for a normal metabolism and, perform redox functions.

Chapter 11

Lignin

> **Abstract.**-The structure of lignin is complex. It consists of a base unit comprising an aromatic ring, a cyclic molecule, and a three-carbon side chain. The deposition of lignin polymers in the cell wall begins with the initiation of thickening of the secondary wall. The lignification of the cell wall proceeds from the region of the primary wall, within the thickening of the secondary wall. The lignin that is deposited changes from lignin type guaiacil to lignin rich in siringil units. Lignin appears to exert a negative effect on the digestibility of cell wall polysaccharides to protect the polysaccharides from enzymatic hydrolysis. The effect of lignin on fiber digestibility is greater in grasses than in legumes. All forages contain lignin, but the concentrations are higher in legumes than in non-legumes, forbs or grasses, in addition, it is higher in stems than in leaves. The total concentration of lignin increases when the maturation of the plant is greater. Lignin also limits the daily amount of dry matter that deer can consume. The microbes in the rumen of the deer are inhibited to break the bonds of cellulose and hemicellulose; therefore, causing a delay in digestion. The energy derived from soluble polysaccharides (such as starch) is also reduced. Resistance to digestion creates a certain stagnation in the digestive tract of deer because microbes have difficulty in breaking the bonds between lignin and structural carbohydrates.

Introduction

Lignin is the main non-polysaccharide component of the cell wall of plants, is considered virtually indigestible for ruminants and is involved in limiting the digestion of cell wall polysaccharides. Lignin is the last of the main biopolymers that has evolved within the kingdom of plants, its most important function in plants is as a structural component that gives strength and solidity to the cell wall and preventing diseases by microorganisms.

These attributes, which are desirable from the perspective of plant survival, serve to limit the nutritional value of plants when consumed by the deer.

Development of lignin

It consists of a base unit comprising an aromatic ring, a cyclic molecule and a three-carbon side chain. Depending on the substitution on the ring, three different monomer units result. In contrast to polysaccharides, lignin has a wide range of bonds, which makes it a very heterogeneous polymer. These can occur between the two rings or between the ring and one of the carbons of the chain or between one chain and another. Lignin is characterized by having an amorphous structure, a high molecular weight, as well as being insoluble in any organic solvent; It gives rigidity and flexibility to vegetables, it is found in the tissue.

The development and growth of the cell wall of plants can be divided into two phases. The growth of the primary wall is the phase when the cell of the plant increases in size through the elongation of the wall. Pectin, xylan and cellulose are deposited during the development of the primary wall, in this phase there is no deposition of lignin. When the elongation of the cell ceases, the cell of the plant changes to allow the thickening of the secondary wall. During this phase, the cell wall progressively swells. The additional polysaccharides are deposited during the development of the secondary wall which is rich in cellulose, xylan and pectin. Ferulic acid is not incorporated into the secondary wall, the deposition of the lignin polymers begins with the initiation of thickening of the secondary wall.

The inclusion of lignin in the cell wall begins in the middle lamina, within the secondary wall. The effect of this is that the polysaccharides most recently deposited in the secondary wall are not lignified. The middle lamina and the region of the primary wall is the most intensely lignified. This may explain why ruminal microbes degrade the cell wall of plants from the lumen to the outside, and because the middle lamina and the primary wall region of lignified cells are never completely digested. The decomposition of lignin during the thickening of the secondary wall is an apparent incorporation of some of the ester-arabinoxylan-ferulate of the primary wall into the ligatures of xylan and lignin.

Lignification

The lignification of the cell wall proceeds from the region of the primary wall, within the thickening of the secondary wall. The lignin that is deposited changes from lignin type guaiacil to lignin rich in siringil units. In conjunction with that deposition of a different type of lignin in later states of lignification, grasses begin to incorporate relatively large amounts of lignin p-coumarate ester into the wall, presumably in the thickening of the secondary wall. The stems of all the forages have a higher concentration of cell wall than the leaves and stems always increase the wall content with maturation. The cell wall of the legumes are rich in pectin and have large amounts of cellulose compared to the xylan that is observed in the pastures. The lignin content of the cell wall of the legumes is greater than that of the grasses, although the magnitude of this difference is increased by the procedure to determine lignin (acid detergent lignin). The change in the composition of lignin is associated with the development of the wall, a change from guayacil to siringyl lignin, apparently the same occurs in all forages. All forage species contain phenolic acids in the cell wall, pastures have higher concentrations than legumes. This difference between grasses and legumes is especially notable for the ester bonds of phenolic acids. The phenolic acids involved in both bonds are very similar in concentration among the forages. Esters of p-coumaric acid appear to be present in all forages, with higher concentrations in grasses than in legumes.

Relationship with the components of the cell wall

Lignin is the component of the cell wall that is recognized as limiting the digestion of cell wall polysaccharides in the rumen. The effect of lignin on the forage digestibility is assumed to have more influence on the digestibility of the cell wall than on the digestibility of the total organic matter of the forage. Lignin appears to exert a negative effect on the digestibility of cell wall polysaccharides to protect polysaccharides from enzymatic hydrolysis. The effect of lignin on fiber digestibility is greater in grasses than in legumes. Recent reports on the negative correlation between lignin concentration and fiber or cell wall digestibility, mention that the lignin composition changes from guayacil lignin to lignins rich in siringyl units with maturation of the forage cell wall; likewise, the digestibility of the mature cell wall is less than that of the immature cell walls, with this it is assumed that the composition of the lignin also affects the digestibility of the cell wall.

Lignin forms part of the cell wall during the formation of the secondary sheet of the cell wall. It is composed of derivatives of phenolic compounds that in turn are synthesized from two routes: that of sinapic acid and acetate-malonate; these two routes form simple monomers to polymers such as lignin and tannins. This close relationship between the biosynthesis of phenols and the normal metabolism of plants is observed in the synthesis of the aromatic amino acids phenylalanine and tyrosine.

The phenolic constituents that have links to the cell wall can be divided into 1) high molecular weight or long chain lignin with covalent linkages with hemicellulose and 2) low molecular weight lignin which are monomers that usually have a single bond with hemicellulose or with another monomer and that could be responsible for some of the crosslinked links between the components of the cell wall. It has been postulated the presence of three types of links between lignin and cell wall carbohydrates in grasses which are: 1) reducible bonds by borohydrate, 2) reducible bonds by alkalis and 3) alkali resistant bonds.

Distribution of lignin in the forage

All forages contain lignin but the concentrations are higher in legumes than in non-legumes, grasses or forbs (Table 11.1), moreover, it is higher in stems than in leaves. The total concentration of lignin increases when the maturation of the plant is greater. The environmental effects on the total lignin content are variable in plants and will be more accentuated in those of rapid growth, but it can highlight the effects of: 1) high temperatures, 2) stress due to lack of water, 3) decrease abrupt light hours (prolonged cloudy), 4) soil fertilization and 5) genetic effects in some plant varieties.

Table 11.1. Lignin content in northeastern Mexico plants consumed by the white-tailed deer

Group of plants	Winter	Spring	Summer	Fall
Legumes	12	12	13	13
Non legumes	10	11	11	10
Forbs	7	6	7	6
Grasses	6	6	6	5

Effects of lignin on digestibility

Several mechanisms have been suggested to explain how lignin affects the digestibility of the cell wall and how some plants that have similar amounts of lignin have different digestibility:

1. *Embedding*; here the lignin that is embedded inside the cell wall inhibits the solubilization of carbohydrates and reduces contact with the cellulolytic microorganisms of the rumen.

2. *Formation of polysaccharide-lignin complexes*; there is an assumption that part of the lignin that is consumed by cattle is solubilized and remains attached to a short chain of carbohydrates (containing approximately 20% of the total weight of carbohydrates). This complex passes without suffering damage by the abomasum and the different.

3. *Specific chemical structure.* In the different varieties of plants due to the variations in the molecular organization that give changes in the condensation of lignin.

4. *Effects on ruminal microorganisms.* It has been shown that phenolic compounds isolated from lignin when present in cultures of microorganisms have the following effects:

    a) They diminish their growth (*Bacteroides succinogenes*).
    b) Decrease its cellulolytic activity (*B. succinogenes, Ruminococcus albus, R flavenfacians*).
    c) They inhibit cellulose attack (*B. succinogenes*).
    d) They inhibit the fungal degradation of the fiber.
    e) They have a depressing effect on the movements of the protozoa of the genus Holotrichis.
    f) They cause a decrease in the production of propionic acid, which causes the bacteria that produce acetic acid to increase.

Lignin vs the deer

Any limitation of fiber digestibility will directly reduce the digestible energy gained by deer. Lignin also limits the daily amount of dry matter that deer can consume. The microbes in the rumen of the deer are inhibited to break the bonds of cellulose and hemicellulose; therefore, causing a delay in digestion. The energy derived from soluble polysaccharides (such as starch) is also reduced. Resistance to digestion creates a certain stagnation in the digestive tract of deer because microbes have difficulty in breaking the bonds between lignin and structural carbohydrates. In general, this contributes to a kind of filling in the digestive tract of the deer that causes a decrease in the consumption of food. Therefore, it is safe to say that deer can consume forages that contain lignin because it does not produce symptoms that show lesions, bleeding noses or death. The deer simply reduces its forage intake. The effects of lignin consumption, therefore, are not to draw attention. However, their production and weight gain if they are affected by the consumption of highly lignified forage.

Chapter 12

Tannins

Abstract.- Due to its importance in animal nutrition, the secondary metabolites of the plants most studied and documented actually are tannins because they are an integral part of the defense system against herbivores and other plant pathogens such as bacteria, fungi, insects and viruses. Also, tannins are an important part of the characteristics that determine the appetite for plants by herbivores due to the astringent characteristics of these compounds. In this way the plant reduces the attack frequency of ruminants and improves their chances of survival. It has been proven that plants that receive greater attack from herbivores are able to increase their concentration of tannins. Generally, tannins are divided into two large groups: hydrolysable tannins and condensed tannins. Low concentrations (2 to 4%) of condensed tannins in plants consumed by white-tailed deer have positive effects on the rumen passage of proteins because they can protect the degradation of plant proteins by ruminal microorganisms. The complex formed by the tannin and the protein would prevent the colonization of the plant particles by the ruminal microflora and its fermentation, possibly contributing to a greater availability of amino acids for the deer, which would be absorbed in the intestine as the tannin-protein complexes dissociate. Likewise, there is sufficient evidence that considers the condensed tannins from intake, as enhancers of productive performance, in animals affected by gastrointestinal parasitosis. In semi-arid regions, deer consume plants with lower amounts of tannins than in tropical regions where plants have elevated tannin content.

Introduction

The white-tailed deer (Odocoileus virginianus) in semi-arid regions composes its diet of a wide variety of individual native plants but must face a great variability in the availability of forage and nutrients throughout the year. However, despite their abundance in pastures, many native species have generally been underestimated, due to insufficient knowledge related to their nutritional potential. Shrub and tree species are an important part of the diet of ruminants in the range, especially in the dry season, when the available herbaceous vegetation is scarce in quality and quantity to meet the requirements of the animals. The deer selects these species, because their foliage has more protein and less fiber than the leaves and stems of the grasses. For this reason, some of these species are degradable and provide protein to supplement the deficiencies of forages of low nutritional quality. Its incorporation in diets of small ruminants is common in developing countries as an energy source, vitamins and minerals.

In the semidesert and desert areas of the world, shrub and tree species are widely distributed and, although they are less preferred by small ruminants, their contribution to their diet is important, due to the ability of this type of animals to consume high amounts of metabolites. secondary plants (MSP) such as tannins, terpenes and essential oils, which may limit deer productivity. Tannins are isolated in vacuoles of plant cells and released into the cytoplasm when the cell is damaged or dies. Condensed tannins (CT) are in free form (soluble), bound to the protein or carbohydrates of the cell wall, but only CT depress the in vitro digestibility of protein and fiber.

Secondary metabolites of plants

It is called primary metabolism of plants to chemical processes that directly intervene in the survival, growth and reproduction of plants. They are chemical processes belonging to the primary metabolism of plants: photosynthesis, cellular respiration, solute transport, translocation, protein synthesis, nutrient assimilation, tissue differentiation, and in general formation of carbohydrates, lipids and proteins that are involved in these processes or they are a structural part of the plants. The primary metabolites of plants (MPP) are the chemical compounds that are involved in the processes: amino acids destined for the formation of proteins, nucleotides, sugars, fatty acids. Due to its universal character in the Kingdom of plants, the processes that intervene in the primary metabolism and its metabolites, are found in all plants without exception.

The concept of MPP was created in contrast to the secondary metabolites of plants (MSP), are chemical compounds synthesized by plants that perform non-essential functions in them, so that their absence is not fatal to the plant, as they do not intervene in the primary metabolism of plants. They are low molecular weight compounds that have great ecological importance because they participate in the processes of adaptation of plants to their environment, such as the establishment of symbiosis with other organisms and the attraction of pollinating insects and dispersers of seeds. The active synthesis of MSP is induced when plants are exposed to adverse conditions such as their consumption by herbivores, attack by microorganisms or exposure to sunlight. In the case of the defense response to pathogens, at least two functions have been reported for MSP: as phytoalexins, that is, compounds toxic to pathogens, or as harvesters of reactive oxygen species.

The MSP intervene in the ecological interactions between the plant and its environment. Most of the MSP were unknown and many of the times were simply known as final products of metabolic processes, without specific function, or directly as waste products of plants. The recognition of biological properties of many secondary metabolites has encouraged the development of this field, for example in the search for new drugs, antibiotics, insecticides and herbicides. In addition, the growing appreciation of the highly diverse biological effects of secondary metabolites has led to reevaluating the different roles they have in plants, especially in the context of ecological interactions.

The MSP are usually grouped according to the chemical substances that constitute them, especially if they have nitrogen in their molecule or not:

1. Phenolic compounds (tannins, phytoestrogens and coumarins).
2. Nitrogenated toxins (alkaloids, cyanogenetic glycosides, glucosinolates, toxic amino acids, lectins and protease inhibitors).
3. Terpenes (sesquiterpene lactones, cardiac glycosides, saponins).
4. Polyacetylene hydrocarbons.
5. Oxalates.

Due to their importance in animal nutrition, the most studied and documented MSP today are tannins.

Tannins

Tannins are phenolic compounds and are considered as a main factor behind the problems of low nutritional value of forage legume plants. They are an integral part of the defense system against herbivores and other plant pathogens such as bacteria, fungi, insects and viruses. Tannins are an important part of the characteristics that determine the appetite for plants by herbivores due to the astringent characteristics of these compounds. In this way the plant reduces the attack frequency of ruminants and improves their chances of survival. It has been proven that plants that receive greater attack from herbivores are able to increase their concentration of tannins.

Tannins are widely distributed in the foliage of plants and, in some herbivores, they decrease the palatability and digestibility of dry matter and protein. Sometimes they act as toxins, rather than as inhibitors of digestion. The diversity of effects of tannins on digestion is partly due to the physiological capacity of the animals to use them and, on the other hand, due to the differences in the chemical reactions of the different types of tannins. In some mammals, salivary proteins react with tannins. In the deer these tannin-binding salivary proteins are glycoproteins containing large amounts of proline, glycine and glutamate / glutamine, but are not related to the tannin-binding salivary protein found in non-ruminant species.

Tannins are substances not well defined chemically, but they are grouped because they have some common properties. They comprise a small part of the broad and diverse group of plant phenolic compounds, which includes phenolic acids of 7 to 9 carbon atoms, such as gallic and p-coumaric acids, flavans of 15 carbon atoms, and lignins. Generally, tannins are divided into two large groups:

1. *The hydrolysable tannins (HT)*. They consist of a central nucleus of carbohydrate to which they are united, through ester bonds, phenolic carboxylic acids. They are esters of sugars of gallic or ellagic acids.
2. *The condensed tannins (CT)*. They consist of oligomers of two or more flavan-3-ols, such as the catechin, epicatechin or the corresponding gallocatechin. They are also described as proanthocyanidins. Depending on the chemical structure of the monomer unit, the number of hydroxyl radicals, they are classified into four groups, the two most common being procyanidins and prodelfinidines.

The large number of phenolic hydroxyl groups that tannins possess makes them very reactive, providing them with numerous anchoring points capable of forming hydrogen bonds, this being the reason why they form reversible associations with other molecules, demonstrating a greater affinity for proteins due to the strong tendency to form hydrogen bonds between the hydroxyl groups of the tannins and the oxygen of the carbonyl group of the peptides. It has been found that tannin-protein complexes are more easily formed at a pH close to 6.0, (average value in the rumen) and that they dissociate at a pH of less than 3.5 and above 8.5. In addition, it has been reported that the union is stronger as time progresses and the more insoluble in water is the tannin molecule. The special characteristics of these tannin-protein interactions mean that condensed tannins have less affinity for the formation of bonds with proteins than hydrolysable tannins, which have greater conformational flexibility of their molecule. Proline-rich, open-structure proteins have high affinity for tannins, while glycoproteins, globular and low-molecular-weight proteins have little affinity.

The HT as CT are mainly found in leaves of trees, shrubs and legumes. The tannin content of trees and shrubs varies widely between species, as well as seasonally and with the phenological state of the plant. Environmental factors greatly alter the concentration of tannins in forages, in general a low intensity of light and low temperature reduces the concentration of tannins, while the drought increases it, consequently a higher concentration of tannins is expected in the middle of summer. The concentration of tannins is also a function of the maturity of the forage, being higher in mature forages.

Effects of tannins on nutrition

Its high degree of reactivity leads to its interaction, both with the proteins of plants, which decreases its accessibility, and with the digestive enzymes of herbivores such as deer, which reduces its use (decrease in the digestibility of organic matter), as well as with the mucoproteins of saliva or directly with taste receptors, which causes the sensation of astringency characteristic of tannins and, consequently, the low palatability of plants that contain high amounts of these compounds.

In living cells of plants, tannins are stored in the vacuoles product of vesicle coalescence of the endoplasmic reticulum. However, in states of senescence and death of the cell, the condensed tannins become part of the cytoplasm in the cells of the cell wall. The condensed tannins contained in the vacuoles of living cells are released by the processing of mastication by the deer, eventually joining the proteins of the diet, the polysaccharides and proteins of the cell wall and remaining part of them as free tannins. In their passage through the deer digestive system, these three fractions undergo different transformations and exchanges between them, being the free tannins the only ones susceptible to undergo degradation or absorption. Once the condensed tannins have passed through the rumen, the gastric (pH 2.5) and pancreatic and biliary secretions (pH 8-9) dissociate the tannin-protein complex, leaving a substantial amount of condensed tannin bound to the cell wall that goes to be excreted in the feces together with lignin, which would contribute to an apparent decrease in the digestibility of the acid detergent fiber from the diet.

Adverse effects of tannins

The biological activity of the CT depends on two main factors: the concentration and the structure of the same. In any case, when the animal consumes a diet with high amounts of tannins, both condensed and hydrolysable, there is a decrease in the consumption of food, as well as the digestibility of nitrogen compounds and animal behavior, can damage the intestines, liver and kidneys. At low concentrations it prevents tympanism and increases the flow of non-ammoniacal nitrogen and essential amino acids of the rumen,

The incidence of tannins in a medium as complex as the rumen, has been observed the inhibition of many microbial activities by their effect: from the reduction of RNA synthesis to the proteolytic or cellulolytic activity due to its action on microbial enzymes. Studies with functional microbial groups have shown that tannins reduce the population of cellulolytic and proteolytic bacteria. However, secondary compounds vary in their structure and concentration in plants; consequently, the mechanisms by which they affect animals in the pasture, such as deer, also vary. Although ruminal microorganisms have metabolic mechanisms that prevent, tolerate and regulate the consumption of toxic compounds, the negative effects of high levels on animal productivity are attributed to the fact that they establish, with cellulose and protein, complexes resistant to ruminal degradation, reduce the adhesion of microorganisms to substrates, microbial colonization, the activity of the 1,4-endoglucanase enzyme, the degradation of the protein, the absorption of amino acids in the intestine, digestibility and consumption.

Likewise, it has been mentioned that forage species with CT levels greater than 5% cause a reduction in dry matter intake, which affects the productivity of the animals. This decrease is due to the effect of TC on palatability, which decreases digestion, the formation of complexes between salivary proteins and tannins which causes a sensation of astringency that can increase salivation, decreasing the palatability of the species. Tannins reduce the rate of fermentation and cause a rumen filling effect, to more severe situations in which the digestion of fiber and nitrogen is reduced, it can also reduce the digestibility of the cells of the wall by adhering to bacterial enzymes or by forming indigestible complexes with structural carbohydrates.

Beneficial effects of tannins

Low concentrations (2 to 4%) of CT in plants consumed by white-tailed deer have positive effects on the rumen passage of proteins, urea recycling and on production and animal health. Tannins can protect the degradation of plant proteins by ruminal microorganisms: the complex formed by tannin and protein would prevent the colonization of plant particles by the ruminal microflora and its fermentation, possibly contributing to a greater availability of amino acids for the deer, which would be absorbed in the intestine as the tannin-protein complexes dissociate. However, some researchers consider it unlikely that tannins will pass through the intestine, with the gradual increase in pH, without interacting with other proteins, whether of the diet or endogenous, with which they wonder if the result in the attempt to protect proteins from degradation in the rumen is a real gain in nutritional value. The depressant action of tannins on the intestinal activity of trypsin and amylase, deduced from their activity in fecal samples, would support these hypotheses.

In addition, in practice, tannins are used to prevent excessive foam formation (tympanism) in ruminants, because they reduce the concentration of proteins in the rumen. In addition, they reduce the incidence of myiasis in sheep (ectoparasites). Likewise, there is sufficient evidence that considers CT from ingestion, as enhancers of productive performance, in animals affected by gastrointestinal parasitosis. That is, their productive level is not affected by the disease. This phenomenon is called Resilience. Fundamentally, in the ovine species, this phenomenon has been studied the most.

Table 11.1 shows some effects in animals consuming plants with different CT levels.

Table. 11.1. Effects of condensed tannins in the diet on digestion

CT in deer usually:
Reduce the ruminal digestion of plant proteins
Reduce concentrations of ruminal amino acids
They reduce the solubility of proteins
Increase the amount of protein that reaches the small intestine of the deer
They inhibit some ruminal bacteria
They reduce the digestibility of nitrogen; increasing the concentration of fecal nitrogen
They reduce the excretion of fecal nitrogen
They reduce the excretion of urinary nitrogen
Decrease or reduce the rate of absorption of amino acids in the small intestine
Reduce the amount of dry matter available to be digested
Reduce the proportion of energy in the diet that is lost in the form of methane
They alter the selection of their diet
Decrease voluntary consumption
Decrease the speed of food digestion
Increase ruminal volume
Decrease the speed and concentration of volatile fatty acids in the rumen
Occasionally they increase or even if they usually reduce the absorption of amino acids the small intestine
Improve tolerance to gastrointestinal parasites such as nematodes
Reduce the number of gastrointestinal parasites such as nematodes
They damage the abomasum o small intestine.

The specific effect of tannins depends on the interaction between the characteristics of the tannins (condensed or hydrolyzable, molecular size, configuration) and the adaptation of the animal to neutralize or metabolize different tannins, the result of the plant-animal interaction requires the understanding of the characteristics of tannins and the physiology, ecology and evolution of the animal. Some herbivores can degrade and absorb a certain amount of tannins such as sheep and cattle. In a study conducted with sheep and white-tailed deer, consuming a pelleted diet with a mixture of alfalfa hay and quebracho tannins (used for the skin cut), reported that sheep excreted around 40% of the consumed tannins, suggesting that 60% was absorbed and metabolized, while the white tail deer feces contained all the amount of tannins consumed, suggesting that the deer did not absorb or metabolize anything.

Most herbivores such as deer that are highly selective of the forages they consume, select plants with relatively low concentrations of tannins. The white-tailed deer in northeastern Mexico prefer to select mature leaves than the regrowths of the *Acacia rigidula* (Table 11.2), apparently the mature material contains less condensed tannins; even when the new leaves contain more nutrients. This has been corroborated that domestic animals such as goats and wild animals such as white-tailed deer counteract the negative effects of secreting tannins, from their enlarged salivary glands, proteins (glycoproteins containing large amounts of proline) tannin binders. When these animals consume plants with high concentrations of tannins, these salivary compounds bind with the tannins making them inactive. The tannin-protein compounds allow the browsers the ability for a high digestion of fiber and protein even when they consumed forages with a high concentration of condensed tannins.

It has been found that goats exhibited a higher level of tolerance to the effect of tannins than sheep, while in sheep the consumption of food, nitrogen balance and the concentration of ammonia in the rumen were depressed by 8%, 159% and 50%, respectively when they were infused intraruminally with tannins. The goats only showed a depression of the ruminal ammonia concentration of 39%, while the excretion of nitrogen in urine was only 17% of the N ingested in goats, in the sheep it was 44%. It has been reported a slight increase in the consumption of dry matter, crude protein and N retention in lambs receiving a diet high in concentrated tannins and inoculated with *Eubacterium cellulosolvens* capable of tolerating 0.5 g/l of condensed tannins, compared to sheep receiving the same diet but inoculated with bacteria from the same culture subjected to autoclave. The goats must have some other detoxification mechanism of the tannins since they do not have, specifically, proline-rich proteins in their saliva. A possible means would be the rumen in which a strain of bacteria *Selenomonas ruminantium* subspecies *ruminantium* has been identified, provided with enzymes with tannin-acylhydrolase activity and, therefore, capable of growing in media with tannic acid or condensed tannins as the sole source of energy.

Tannins in the plants that the deer consumes

The 32 species of native shrub plants of northeastern Mexico that are part of the diet of white-tailed deer (Table 11.2) have an annual average of 4% condensed tannins; Seasonally the average concentrations were very similar 3, 4, 4, and 4% in winter, spring, summer and autumn, respectively. Even though the largest number of plants contain tannin levels between 4 and 5%, species such as *Acacia berlandieri* (23%, annual average), *Acacia rigidula* (19%, Figure 4), *Ziziphus obtusifolia* (13%), *Cercidium macrum* (9 %), *Desmanthus virgathus* (9%) and *Leucaena leucocephala* (8%) contain high concentrations of condensed tannins. Of these six native shrubs, in order of importance, the most selected by the white-tailed deer are *A. rigidula, C. macrum* and *A. berlandieri*. It should be mentioned that more than 40% of the monthly diet of white-tailed deer in the northeast of Mexico is composed of *A. rigidula*.

Apparently, deer in northeastern Mexico and south Texas, USA consume diets, during most of the year, mainly composed of shrub species, some contain high amounts of tannins, others have intermediate levels and the least are practically devoid of tannins. When correlating the content of condensed tannins, which are found in the shrub plants shown in Table 11.2 and the digestibility values of the organic matter of these same plants, a negative relationship was found ($r = -0.53$, $P < 0.001$) . This could mean that shrubs with higher tannin content, organic matter is less digested by rumen microbes.

When it was evaluated the content of TC in the foliage of trees and shrubs of the Tierra Caliente Region of Michoacán, Mexico (Table 11.3) reported levels of 5.3% in 57 plants evaluated; of which 21 were from the legume family and the rest from other families. It should be noted that in this tropical region of Michoacán, Mexico, the plants of the legume family contain less tannin than the other families. In any case, the reported average levels (5.3%) can be considered, in the deer when consuming this type of plants, that could have negative effects on their productive performance. *Lysiloma acapulcensis* was the plant with the highest tannin content. However, beneficial effects could be caused by the plants that the deer consumes in northern Mexico with values around 2.1% (Table 11.4). In this study, carried out in a semi-arid zone, the concentration of tannins is higher in the plants of the legume family (3.9%) compared to those of other families

(2.1%). The plant with the highest tannin content was *Calliandra eriophylla*. A similar trend is shown in Table 11.2, where the legume had values of 6% and the other families had only 1.9%. Therefore, based on these three studies, compared to other families, plants of the legume family in tropical climates can develop a lower content of condensed tannins than those legume of arid and semi-arid climates.

Anti-parasitic activity of tannins

Recently, evidence has been found of the nematicidal effect that tannins may have on gastrointestinal nematodes (GIN). Tannin plants may have a direct antiparasitic activity but could also have an indirect effect through improving the immune response of animals against GIN. Recently, several studies have been reported in which tannins can be shown to improve resilience (less clinical signs, better growth and wool production) and resistance (lower number of nematode eggs in feces, lower parasitic load and lower fertility of females parasitic) of goats and sheep infected with GIN. This has been demonstrated using pastures with tannins, hay from tannin plants and forage from trees or woody plants from different latitudes.

In Yucatan, Mexico it has been discovered that the foliar acetonic extract of various native tannin trees has an anthelmintic effect in vitro against *Haemonchus contortus*. Apparently, the mechanism of action of tannins on the infecting larvae of *H. contortus* and *Trichostrongylus colubriformis* is to prevent these parasites unsheathe. This avoids that the GIN can be established in their site of action and can continue with their evolutionary cycle. The mechanism of action of tannins in adult nematodes is also studied. The tannins seem to have a different mechanism of action in the latter. Apparently, the tannins join the mouth and possibly the reproductive system of the parasites (due to the affinity of the tannins to the proline-rich proteins of the cuticle of the nematode).

Table 11.2. Seasonal variation of condensed tannin content (% dry base) in the forage of native shrubs that grow in northeastern Mexico and southern Texas, USA

Scientific name	Winter	Spring	Summer	Fall	Mean
Legume					
Acacia berlandieri (L.) Wild.	17.0	33.0	24.0	19.0	23.3
Acacia farnesiana (L.) Wild.	2.0	2.0	3.0	2.0	2.3
Acacia rigidula Benth.	15.0	22.0	22.0	18.0	19.3
Acacia wrightii Benth.	1.0	1.0	1.0	1.0	1.0
Caesalpinia mexicana A. Gray.	1.0	1.0	1.0	1.0	1.0
Cercidium macrum I.M. Johnst.	8.0	8.0	10.0	8.0	8.5
Desmanthus virgathus L.	7.0	9.0	7.0	12.0	8.8
Eysenhardtia polystachya (Ortega)	2.0	1.0	1.0	1.0	1.3
Leucaena leucocephala L.	8.0	7.0	6.0	10.0	7.8
Parkinsonia aculeata L.	0.0	0.0	0.0	0.0	0.0
Pithecellobium ebano (Berl) Muller.	3.0	2.0	3.0	2.0	2.5
Pithecellobium pallens (Benth) Standl.	1.0	2.0	1.0	2.0	1.5
Prosopis glandulosa Torr.	1.0	1.0	1.0	1.0	1.0
Mean	5.1	6.8	6.2	5.9	6.0
Non legume					
Bernardia myricaefolia (Scheele) Wats.	0.0	0.0	0.0	1.0	0.3
Bumelia celastrina H.B.K.	4.0	2.0	2.0	3.0	2.8
Castela texana T. and G. Rose.	3.0	4.0	5.0	3.0	3.8
Celtis pallida Torr.	0.0	0.0	0.0	0.0	0.0
Condalia obovata Hook.	2.0	1.0	1.0	1.0	1.3
Cordia boissieri A. DC.	0.0	0.0	0.0	0.0	0.0
Diospyros texana Scheele.	2.0	2.0	2.0	2.0	2.0
Forestiera angustifolia Torr.	0.0	0.0	0.0	0.0	0.0
Gymnosperma glutinosum (Spreng) Less.	4.0	4.0	5.0	5.0	4.5
Helieta parvifolia (Gray) Benth.	0.0	0.0	0.0	0.0	0.0
Karwinskia humboldtiana (R and S) Zucc.	3.0	3.0	3.0	2.0	2.8
Larrea tridentata DC.	1.0	1.0	2.0	2.0	1.5
Leucophyllum texanum Berl.	1.0	1.0	2.0	1.0	1.3
Opuntia engelmannii Engelm.	0.0	0.0	0.0	0.0	0.0
Porlieria angustifolia Engelm.	2.0	1.0	2.0	1.0	1.5
Schaefferia cuneifolia Gray.	0.0	0.0	0.0	0.0	0.0
Zanthoxylum fagara (L.) Sarg.	0.0	0.0	0.0	0.0	0.0
Ziziphus obtusifolia T and G.	13.0	14.0	11.0	14.0	13.0
Mean	1.9	1.8	1.9	1.9	1.9

Table 11.3. Content of condensed tannins in forage species from Michoacán, Mexico

Scientific name	Tannins, %	Scientific name	Tannins, %
Legume		Non legume	
Acacia acatlensis	3.20	*Amphipterygium adtringens*	6.99
Acacia macilenta	1.91	*Bumelia socorrensis*	40.80
Andira inermis	1.28	*Buncosia* sp.	3.20
Andira sp.	1.39	*Bursera heteresthes*	8.08
Bauhinia ungalata	4.50	*Capparis indica*	0.00
Caesalpinia coriaria	5.25	*Cochlospermum vitifolium*	10.06
Caesalpinia platyloba	12.25	*Combretum farinosum*	3.30
Dalbergia congestiflora	0.14	*Comocladia engleriana*	0.10
Diphysa minutifolia	1.89	*Cordia elaeagnoides*	0.46
Enterolobium cyclocarpum	2.90	*Cordia* sp.	0.08
Haematoxylon brasiletto	6.04	*Crescentia alata*	0.42
Leucaena leucocephala	6.58	*Erythroxylon compactum*	1.27
Lysiloma acapulcensis	23.28	*Exostema caribaeum*	0.00
Lysiloma divaricata	6.47	*Ficus cotinifolia*	7.10
Lysiloma tergeminum	3.00	*Ficus* sp.	1.11
Pithecellobium acatlense	1.41	*Guazuma ulmifolia*	1.53
Pithecellobium dulce	1.47	*Gyrocarpus jatrophifolius*	0.32
Platymiscium lasiocarpum	4.50	*Heliocarpus velutinus*	24.28
Prosopis laevigata	2.00	*Jacaratia mexicana*	0.00
Pterocarpus arbiculatus	0.46	*Mangifera indica*	2.70
Senna skinneri	2.50	*Mastichodendron capiri*	0.61
Mean	4.40	*Parmentiera aculeata*	0.00
		Pseudobombax ellipticum	4.00
		Psidium guajava	30.96
		Randia echinocarpa	1.06
		Randia sp.	0.00
		Randia watsoni	0.70
		Simira mexicana	7.28
		Spondias purpurea	1.96
		Stemmadenia obovota	0.00
		Swietenia humilis	5.05
		Tabebuina palmeri	0.00
		Thevetia ovata	0.00
		Vitex hemsleyi	0.00
		Vitex mollis	2.30
		Xylosma sp.	45.38
		Ziziphus amole	1.17
		Mean	5.74

Table 11.4. Content of condensed tannins in forage species of Durango, Mexico

Scientific name	Tannins, %	Scientific name	Tannins, %
Legume		Non legume	
Calliandra eriophylla	9.1	*Quercus eduardii*	0.4
Acacia constricta	3.0	*Quercus grisea*	1.0
Mimosa biuncifera	7.4	*Atriplex canescens*	0.3
Cassia wislizeni	1.4	*Flourencia cernua*	0.3
Acacia shaffneri	4.4	*Celtis pallida*	1.2
Prosopis leavigata	1.6	*Condalia lycioides*	5.0
Dalea bicolor	4.4	*Larea tridentata*	7.2
Cassia greggii	0.2	*Cordia parvifolia*	2.0
Mean	3.9	*Jatropha dioica*	3.8
		Parthenium incanum	0.2
		Mean	2.1

Chapter 13

Voluntary intake

> **Abstract.-** The feed intakr of white-tailed deer in free grazing shows an evident seasonality, being higher in summer and lower in winter, with fluctuations corresponding to body weight and daily weight gain; males and females show this behavior to a greater or lesser degree. Determine the amount ingested and the quality of the diet of the deer in free grazing, represents one of the most difficult tasks in the investigation because it requires a lot of work, a high cost and, in certain circumstances, it is difficult to obtain precision and accuracy in measurements. This has discouraged the efforts made by pasture and wild fauna researchers to estimate the consumption of deer. As a result, estimates of nutritional status and potential needs are only made to supplement the deer.

Introduction

The deer can consume and digest anything within a range of foods used by the most representative domestic ruminants. In comparative studies of digestibility with deer, it has been found that incompletely digest diets based on straw of low quality, but if they digest more successfully and to a greater degree the diets obtained from browse, compared with domestic species such as sheep. The above can be explained due to the short retention time and the rapid ruminal passage rate observed in the deer. Smaller deer species which normally select low fiber diets do not have the omasaum capacity necessary to process gross particles in low digestibility diets such as mature fodder or hay. Seasonal changes in feed intake are, therefore, associated with seasonal changes in ruminal capacity without seasonal alteration of the apparent digestibility or the retention time of the food in the deer. At digestibilities lower than 65%, the deer faces the problem of consuming more forage but does not have the capacity to increase its digestibility. The protein content is a good indicator of digestibility. This is due, in part, to the correlation between protein consumption and the state of deer growth. The new weight increase is due to the increase in the consumption of protein and highly digestible carbohydrates. However, the protein can also be a limiting factor for microbial growth and therefore for the fermentative capacity of the rumen. The deer selects a variety of diets to capitalize a nutrient balance or avoid phytotoxins.

Need for nutrients

Herbivores ingest food to satisfy the need and desire for nutrients, with energy and protein being the most limiting. For a female, the demands of maintenance, lactation, growth and conception add up to the total of her requirements. The energy required to increase if the deer needs to travel long distances or for thermoregulation. The voluminous density of the feed, passage rate and digestion in the rumen of the deer interact with the needs and desires of food ingestion. The deer stops consuming food when it obtains a maximum and optimum intake of nutrients; which occurs under ideal conditions and when nutritional needs are met, anatomical sensors send a signal to the brain for satiety to occur. Low dry matter consumption, whether produced by poor quality and quantity of forage, is the main impact of inadequate nutrition of grazing animals such as deer.

Selective behavior of the ruminant

It is the first line of defense against the negative effects of plants that contain toxic compounds. The grazing animals are the most sensitive to the quality and anti-nutritive compounds of the plants. For example, animals select high quality diets from the available forage. They also select plants and plant parts of relatively low toxicity. The animals accompany these wise decisions relating the taste of the plants with the positive or negative digestive consequences. The chemical and structural compounds of plants dictate the potential digestible energy, production of nutrients, or toxicity of a plant. The digestion and detoxification skills of grazing animals, and their ruminal microbes, determine the production of nutrients, energy, or toxins from plants. The results of these interactions between the plant and the animal determine the palatability of the forage. The key to how animals respond to anti-quality factors in plants is therefore focused on the consequences of intake.

When a grazing animal smells and tastes a plant, the taste is either pleasant or unpleasant, depending on previous experiences of grazing. When a plant is consumed, it provides a feedback effect during digestion. If the consumption of a plant improves the status of the animal's nutrients or energy, the taste of the plant becomes more desirable and pleasurable. If the consumption of the plant causes damage, the taste becomes adverse and unpleasant. These relationships of taste consequences form the basis of tastes and dislikes, and therefore, the goat looks for highly palatable foods and avoids non-palatable ones. The resulting patterns of behavior generally lead to an increase in the consumption of nutritious foods and limit the consumption of toxic or low-quality plants.

Ruminants that live in multigenerational groups in which dietary information can easily be passed from more experienced animals to non-experienced ones. Young animals, therefore, do not require complete and perfect information about the diet at birth. Learning from the mother and can be started before young animals consume their first foods. Flavors in the uterine fluid and mother's milk can influence the preference of foods. However, as animals mature, they are more influenced by their own dietary experiences than by those of their mother or other social models.

Certain ruminants such as goats and cervids have adaptive patterns of intake so they have a strong and natural tendency to select diets composed of various types of plants and test the available plants on a regular basis. This behavior can increase the possibility of ingesting the necessary nutrients and reduce the potential to consume excess toxins. The toxic effects of a plant are determined mainly by the amount consumed, although the rate of ingestion is also important. Grazing animals can avoid toxicosis by limiting their intake of a particular toxic plant each day to allow sufficient detoxification time, and limit the cumulative potential effects of a particular toxin.

Selectivity indices

Determining what a herbivorous animal consumes under grazing conditions is not an easy task; because the final diet selected by the animal corresponds to a particular situation and is a function of many factors related between the plant and the animal. The individual characteristics of plants play an important role and greatly influence their acceptance or rejection. These characteristics determine the palatability of plants. However, there are also unique characteristics; morphological and physiological and behavior of the animal that interact to determine the feeding strategy of each species, or how to start exploiting the available food source. The joint effect of this, is manifested in the feeding behavior called selectivity, which is a response of the animal to choose one or more species of plants (and part of the plant).

The selectivity of grazing ruminants for a specific plant species is determined, in part, genetically, by previous experience or conditioning, by the physiological status and nutritional status and partly by the circumstances prevailing in the environment, including availability relative of some plants, among which the selection is carried out. Sensory thresholds for certain chemical substances, is one of the most important genetic characteristics that has been determined to influence the dietary selection of goats. This coincides with recent theories about the evolution of chemical defenses of plants. Such secondary compounds such as alkaloids and tannins tend to evoke acidic sensations. In the case of tannins, however, there is uncertainty that if the sensation is an acid taste or the result of the astringent action on the mucosal epithelium of the mouth, which can be intercepted as a touch, instead of a gustative sensation.

Several morphological factors related to grazing behavior apparently contribute to the successful adaptation of small ruminants to a wide variety of environmental conditions. These include a movable upper lip and the ability to assume a bipedal position in food intake. Among the domestic species of ruminants, only the goats seem to have this last quality; however, there is no information available, which has been specifically tested on whether the bipedal position would give the goat a competitive advantage. Intuition tells us that the ability to browse above the head would increase the biomass of available forage in wooded areas and in scrub areas grazing the species above the head that are palatable. Probably the greatest advantage will occur during periods of drought, where the forage layer of the soil is dry or damaged by grazing. The tendency of some deep-rooted trees and certain shrubs to maintain persistent leaves during periods of drought seems to offer a nutritional advantage to an herbivore, such as the goat, which can use such forage. Specific research is required to document this hypothesis and to quantify the advantages; in case there are. A situation where goats and sheep graze at the same time, would be appropriate for such experiments.

Generalizing on studies of selectivity in diets, is very difficult since practically all have been carried out under different conditions of plant availability. Consequently, the results tend to be from specific places and offer little if any basis for formulating dietary selection principles. Despite such limitations, these studies, when applied to the place from where the data originated, have provided grassland managers with a partial basis for management decisions and environmentalists as a first approximation to explain how the relationships of the communities of plants by grazing.

The presence of a species varies according to the time of use, animal function, changes in the structure of the plant and many other factors related to animals and plants, while the interactions of these factors of preference are important in determining of the plant species to be consumed by the goats; Many of them are not suitable for inclusion as part of the preference index, since the values of the preference indexes should interpret and integrate numerical values with other influencing factors not included in the index.

The measures of the frequency of plant species, both in the diet and in the pasture, are useful to interpret the preference to forage. The frequency in the diet assesses and measures the food consumed, while the frequency in the pasture assesses and measures the distribution of the plants within a certain community. These values, when incorporated into a relative index of preference, increase its sensitivity, but it is not substitutable for the measurement of the composition of the diet or for the availability of forage in the rangeland.

The values of a preference index for a given species can indicate either; if it was preferred or rejected. But the primary value of the preference indexes is to value several plants, taking into account their palatability under certain specific circumstances. An index of 1.0 indicates that the percentage of a species in the diet is equal to the percentage of that species available in the range. Values above or below this index indicate a selectivity or evasion, respectively.

Palatability

The biggest problem in evaluating the consumption of food is that the animal rejects the food, one reason may be palatability, which is defined as the pleasure to taste the food. The cause of rejection, which may be due to taste or other physiological reactions, is hardly recognizable. The concept of palatability can also be referred to the consumption of free access to food on some fractions of the diet, so offering enough food so that the animal can select it can be recommended. The selection of a forage presumes morphological and nutritive differentiations in the plants, generally a hungry animal is very little selective.

Appetite

If the available foods are acceptable, the amount of food consumed per unit of time is a function of appetite. The consumption of food is regulated physiologically in the short and long term, otherwise, malnutrition or obesity could occur. The centers of satiety and hunger are found in the hypothalamus. In ruminants, the volatile fatty acids, products of microbial fermentation carried out in the reticulum-rumen are the main source of energy. Intraruminal infusions of acetate, propionate or butyrate (or mixtures of these acids) decrease feed intake. However, this effect may not be a direct action in the appetite control center of the hypothalamus because intravenous infusions produce decreased appetite month, so that there may be other centers of appetite control or other metabolites are more important in regulating the hypothalamic function. The white tail deer consumes food based on the energy density of the diet. When the deer consumes diets with digestibilities less than 50%, there are physical limitations in the digestive tract that limit the intake of dry matter to less than the requirements for maintenance and the fawns can lose weight.

Food intake

It is one of the best regulated homeostatic mechanisms of the animal organism. The regulation of consumption occurs at different levels. For example, the animal must balance the acquisition of nutrients to cover the daily and seasonal metabolic demands by regulating the incidence and consumption between meals. Such a system must monitor several environmental conditions (photoperiod and food availability, ruminal filling and absorption of nutrients, body fat and body energy and nutrient needs). These monitoring and control mechanisms must be incorporated hierarchically to allow the maintenance of the energy balance in different nutritional environmentals.

Generally, animals consume food to provide their own tissues with nutrients that are required for the processes of body maintenance, growth (fat deposition in adult animals), milk production and work. However, due to the variety of ingredients that make up a diet and that are consumed by the animal, it is not likely that the composition of nutrients provided could accurately meet the proportion of nutrients required by the animal. Therefore, providing the exact requirement of one nutrient can result in the deficiency or excess of another. The animal stops eating due to physical or metabolic limitations, in this way the animal has to decide to what extent the disadvantages or deficiencies or excesses of certain nutrients weigh more than the advantages of trying to satisfy the animal's energy requirements, which, it is believed that they are the driving forces of the intake.

The factors that affect food consumption in ruminants are several and are classified as follows:

Physical factors.- These are the ones that generally have the greatest influence on the increase in the volume of the stomach, occupied by the ingestion of a food, and the degree to which the volume is decreased due to the digestion and passage of the intake. The fibrous content of the cell walls is the main factor in this respect, because these structures are less soluble and take more space in the rumen, than the cellular content. Pastures contain a large proportion of their organic matter (35-80%) composed of cell walls, which provide the structural integrity of plants. Structural carbohydrates (cellulose, hemicellulose and pectins) are degraded by ruminal microbes, which enables the ruminant to use a source of energy that is not efficiently used by non-ruminants. The distribution of the different molecules within the plant and their links between them are important factors that affect the ability with which microorganisms can degrade cells and, therefore, the space occupied in the digestive tract. In addition, the physical characteristics of the cell wall or the same fibrous particles as the origin of tissue, shape, flotation and specific gravity affect the speed at which the particles are degraded and their ease of passage.

Spray resistance (reduction of particle size) is positively correlated with the fiber content; however, the ratio between the fiber determined using the neutral detergent solution (NDF) and dry matter intake (DMI) is not always constant. A significant effect of different types of forages (C3 or C4 grasses or legumes) on the slope and intercept of the regressions between DMI and NDF in cattle and sheep has been demonstrated, indicating that the effect of ruminal filling using the NDF may vary depending on the type of forage. This can be explained by the differences in molecular distribution of the different structural polysaccharides. On the other hand, groups of forages with similar DM digestibility, the fiber content is higher in legumes compared with temperate climate grasses purchased with those of warm climate and leaves compared with stems. In addition, the low digestibility of tropical grasses compared to those of temperate or legume climate reflects an interlacing and, therefore, a rigid cellular structure. Likewise, the greater digestibility of legumes compared to grasses may be due to the length of their leaves. Pasture particles are inherently more voluminous and lazy, with low specific functional gravity (SFG) and are easily interlaced, while legume bites are short and high in SFG and therefore disappear from the rumen very quickly. Therefore, the potential consumption is dependent not only on the fiber content, but also on the original structure of the plant and the way it is degraded during digestion.

The dry matter content of foods can also influence the space occupied within the digestive tract. The pre-drying of the pastures before ensiling them, consistently, has shown that the ensilages are consumed in greater quantity compared with the pastures that have not been previously dried and in some occasions the consumption increases up to 44%. One explanation is that the effectiveness of the bite during the consumption and therefore the speed of the breaking of the particles was improved with the dehydrated material. However, the advantages in terms of animal production have been less evident in recent years because as the control of the fermentation in the silos has improved, and the pastures have been harvested in states of premature maturity, giving materials with higher digestibility, by which the effects of dehydration are not very effective. The practice of dehydration continues due to the strategy of reducing runoff rather than to improve the nutritional quality.

Physiological factors.- A large number of neurotransmitters (norepinephrine, epinephrine, dopamine, serotonin, and gamma-aminobutyric acid) and hormones (insulin, cholecystokinin, neurotensin, glucagon, calcitonin, and growth hormone releasing factor) control the balance of nutrients and food consumption. These compounds stimulate several brain sites (hypothalamus, cerebral cortex, autonomic nervous system, cerebellum) that regulate consumption. A decrease in consumption due to pregnancy has been observed, which is consistent with previous results reported in the literature and proposed that hormones are involved in the regulation of consumption, in addition to the physical limitations resulting from the increase in the size of the the fetuses diminishing space in the rumen.

Structure of the rangeland.- In grazing animals, the distribution of plants can restrict, not only because of the space of the food that takes place in the digestive tract, but also because of the limited amount of forage that the animal can take in at 24 hour period. Characteristics such as density and height of plants, can influence their capacity, either by the capacity of apprehension and, therefore, bite size, which has been shown to be the main factor in the daily consumption of food. Therefore, in grazing animals not only the structure of the plant, but also the characteristics of the pasture should be considered to predict the food.

Metabolic factors.- Not all differences in consumption between pastures can be explained by factors that can influence the space occupied in the digestive tract. For example, it is generally accepted that the storage of pastures results in a decrease in consumption. However, after improvements in silage storage and preservation techniques, low silage consumption compared to hay is considered a minor problem. Organic acids and nitrogenous substances such as amines, produced during the silage process and the additives used to improve the nutrient quality of the silo, have sometimes been implicated as responsible for the decrease in consumption observed when fresh, freshly harvested grass, It is silage. Ruminal infusions of the individual constituents or mixtures of them have indicated that there is no constituent that is responsible for the low intake, but the additive effects of a large number of substances may be responsible.

Size of the animal.- The size of the animal is the factor most correlated with the intake of food. Large animals consume larger amounts of forage; however, the relationship is not isometric, although using allometric scales with muscle mass, consumption is commonly expressed based on the metabolic weight of the animal (live weight$^{0.75}$). Although it is not surprising, either, that parameters related to the processing of fiber are also related to the size of the animal. It has been found that the rumination time per gram of NDF decreased exponentially with the live weight of the animal. Likewise, it was demonstrated that the time required to break the large particles of the fiber into smaller particles, have a smaller scale in the retention time with a live weight$^{0.27}$, while the passage of the digest towards the lower digestive tract it is isometric with live weight. The allometric relationships have a great influence on the variations attributable to the body mass and could be incorporated into models to improve the prediction of consumption. Young growing animals apparently can consume more food than adult animals with the same weight and intake seems to be allometrically related to live weight$^{0.60}$, instead of live weight$^{0.75}$, the latter is assumed to correspond to adult animals. This difference reflects the effect of the metabolic rate and, therefore, the physiological state of the animal over intake.

Grazing animals.- The animal factor can also influence consumption during grazing. Consumption in grazing animals is dependent on the degree of ingestion (size and number of bites) and the time of grazing. Although the structure of the pasture and the amount of forage available affect these parameters, as previously described, the animals are able, within certain limits, to alter the behavior during grazing, increasing the grazing time where the pasture is scarce and, therefore, grazing activity increases. The effect of physiological status on grazing intake, has also proven; For example, when forage availability is high, consumption increases up to 460 g organic matter kg^{-1}, increases in milk production have also been observed. The effects have been observed in the rate of intake and/or time of grazing.

Level of intake.- The size and number of bites are not independent. For a given forage a given time of mastication is required, an increase in the size of the bite will cause an increase in the time of mastication, a decrease in the number of bites resulting in equal rates of intake. The fact that animals can alter their rate of consumption has been shown when animals increase their intake after a prolonged fast. This observation has shown that animals rarely consume food up to a maximum limit, although there have been some indications that animals increase their level of intake when their grazing time has been restricted after they learned that they will only be allowed to graze for a limited

Time used for grazing.- When the structure of the plant limits the size of the bite and, therefore, the degree of intake, the time dedicated to intake can be altered, to compensate for the decrease in the size of the bite. However, there seems to be a maximum period of grazing that the ruminant uses to carry out the intake of food. Sheep and cattle graze from 13 to 15 hpurs day^{-1}, while ruminants consume food up to a maximum of 12 h d^{-1}. Therefore, if the size of the bite is less than a certain limit, the animals will not be able to achieve their maximum filling capacity. This occurs because of the oral time of processing the food reaches a maximum limit, which involves the grasping, chewing and rumination. However, this limit, if any, is dependent on the physiological state, because the increase in milk production results in an increase in grazing time.

Adverse experiences during grazing. - Ruminants reduce the ingestion of food in response to an indisposition resulting from the intake of toxins, such as alkaloids, condensed tannins and glucosinolates and low quality silages. The aversion to a food increases depending on the severity of the damage and decreases as the space of ingestion increases and the damage caused by the food increases. The aversion to food decreases as time passes and the recuperative process counteracts aversion, resulting in a cyclical consumption of nutritious foods containing toxins. A marked decrease is followed by a gradual increase. Similar responses were observed in animals that consumed grains, which appears to be the result of problems caused by organic acids produced during starch digestion.

Even when the presence of toxic foods in places where ruminants have free access to their food, there are few cases of toxicity, since these animals are able to limit the consumption of nutritious plants that contain tannins, depending on the concentration of the toxin. Animals can recognize foods based on previous experiences. Thus, the preference for a food depends not only on the taste, but also on the consequences of such food in the rumen ecology.

Interactions between the components of the diet. - Supplementation can be considered as a measure to increase the supply of nutrients, for animals that are not able to consume the necessary nutrients in forages due to the physical limitations of ruminal filling. Supplementation generally has a positive effect on dry matter intake but may have positive or negative effects on the intake of the basal diet.

Negative effects on the DMI. - A substitution effect is usually manifested when the intake of fibrous foods decreases and varies to the same degree that the digestibility of the base diet varies, among other factors. Under certain circumstances, the effect can be explained according to the simple rules of additivity, where the fiber is replaced 1:1; however, this cannot be considered as a generalized explanation. The degree of substitution in dairy cattle ranges from 0.4 to 0.8 with an approximate average of 0.5. A linear effect has been found by increasing the level of substitution. In addition, an increase in the degree of substitution has been reported as the level of supplementation increases, in addition to a positive relationship in the quality of the forage.

Supplements rich in rapidly fermentable carbohydrates may have a greater inhibitory effect on fiber consumption compared with slowly fermentable supplements, due to the decrease in the digestibility of fibrous forages. Rapid fermentation results in an inhibition of cellulolysis, which has been attributed to the decrease in pH or even an inhibitory feedback mechanism of key digestive enzymes

Positive effects in the DMI. - Under certain circumstances, supplementation can be used to increase the consumption of poor nutritional quality pastures providing a limiting nutrient. In pasture-based diets, the degree of microbial fermentation decreases if the ruminal ammonia content decreases below 50 ml N l^{-1}. Likewise, foods with a crude protein (CP) content lower than 62 g CP kg^{-1} of the DM, the digestion of the fiber is inhibited. In cases where the ammoniacal N concentration limits the microbial fermentation, the supply of N to the microorganisms increases the digestion of the OM in the rumen, which leads to an increase in the degradation and rate of passage of the pastures of low nutritional quality, by means of the removal of physical obstacles allowing the animal that the animal consumes more food. Add small amounts of green foliage to diets made from straw of low nutritional quality can increase the DMI. This could be due to the provision of a source of highly colonized fiber to plant bacteria to the less digestible fiber or through mechanisms that increase the concentration of ammonia above the critical level.

Chapter 14

References

Brown, R.D. 1994. The nutrients for survival and growth. In: Deer. Gerlach, D.S., S. Atwater and J. Schnell (eds.). Stackpole Books, Mechanisburg, PA. EUA, p. 384.

Martínez, S.J., R.M. Pedraza y Y. García. 2001. Influencia del método de secado del follaje y el solvente de extracción en la cuantificación de polifenoles extractables totales. Pastos y Forrajes 24:353-356.

Aastum, MI. 2006. Forage selection by cattle in heterogeneous pastures in Nicaragua. Thesis Mag. Sc. Trondheim, NO, Norwegian University of Science and Technology. 43 p.

Akerman, A.B. y D.G. Johnson, 1991. Gramíneas de Sonora. SARH, COTECOCA, Gobierno del Estado de Sonora, Secretaría de Fomento Ganadero, Hermosillo, Son. México.

Akin, D.E., and A. Chesson. 1990. Lignification as the major factor limiting forage feeding value especially in warm conditions. p. 1753-1760. In Proc. XVI int. Grassland Congr., Vol. III. Nice, France, 4-11 October 1989, Association Francaise pour la production Fourragere, Versailles, France.

Alanís-Flores, G.J., Cano y Cano, G. y Rovalo Merino, M. 1996. Vegetación y Flora del Estado de Nuevo León. Una Guía Botánico-Ecológica, Impresora Monterrey, S.A. de C.V. p. 94

Aman, P. 1993. Composition and structure of cell wall polysaccharides in forages. In: H.G. Jung, D.R. Buxton, R.D. Hatfield, and J. Ralph (Ed.) Forage Cell Wall Structure and Digestibility. p 183. ASA-CSSA-SSSA, Madison, WI.

Ammar, H., S. López, J.S., González. 2005. Assessment of the digestibility of some Mediterranean shrubs by in vitro techniques. Anim. Feed Sci. Technol. 119:323-331.

Annison, E.F. y D. Lewis. 1981. El Metabolismo en el Rumen. Editorial Unión Tipográfica Editorial. Hispano-Americana, S.A. de C.V., México, pp. 10-50.

AOAC, 1997. Official Methods of Analysis 17th Edn. Association of Agricultural Chemists, Washington, DC.

Aranda, M. 2000. Huellas y otros rastros de mamíferos grandes y medianos de México. Instituto de Ecología, A.C. Xalapa, México. 212 p.p.

ARS, 1999. Agricultural Research Service, U.S. Department of Agriculture, USDA Nutrient Database for Standard Reference, Release 13. Nutrient Data Laboratory Home Page, http://www.nal.usda.gov/fnic/foodcomp.

Arthun, D., J.L. Holechek, J.D. Wallace, M.L. Galyean, and M. Cardens. 1992. Forb and shrub effects on ruminal fermentation in cattle. Journal of Range Management, 45:519-522.

Asleson, A.M., Hellgren, E.C. y Varner, L.W. 1996. Nitrogen Requirements for Antler Growth and Maintenance in White-Tailed Deer. Journal of Wildlife Management. 60: 744-752.

Asplund, J.M. 1994. Principles of Protein Nutrition of Ruminants. CRC Press. Boca Raton. EUA. pp. 3-28.

Asplund, J.M. 2000. Structure and Function of the Ruminant Digestive Tract. En: Principles of Protein Nutrition of Ruminants. J.M. Asplund (Editor). CRC Press, Boca Raton, FL, EUA, pp. 5-28.

Atasoglu, C. Wallace, R.J., 2003. Metabolism and de novo synthesis of amino acids by rumen microbes. En: J.P.F. D'Mello. Amino Acids in Animal Nutrition. 2ª edición. CABI Publishing, Wallingfod Oxon, OX10 8DE, Reino Unido. pp. 291-308.

Baba, A.S.H., Castro, F.B., Ørskov, E.R., 2002. Partitioning of energy and degradability of browse plants in vitro and the implications of blocking the effects of tannin by the addition of polyethylene glycol. Small Rumin. Res, 95, 93-104.

Bacic, A., P.J. Harris, and B.A. Stone. 1988. Structure and function of plant cell walls. In: J. Preiss (Ed.) Biochemistry of plants. Vol. 14. Carbohydrates. p 529. Academic Press, San Diego, CA.

Bahnak, B. R.; J. C. Holland, L. J. Verme, y J. J. Ozoga. 1981. Seasonal and nutritional influences on growth hormone and thyroid activity white-tailed deer. J. Wildl. Manage. 43:454-460.

Baiza-Gutman, L.A. y Martínez-Hernández, G. 2007. Estructura y función de aminoácidos, péptidos y proteínas. En: J.J. Hicks-Gómez (editor). Bioquímica. McGraw-Hill. pp. 53-68.

Baiza-Gutman, L.A. y Martínez-Hernpandez, G. 2007. Estructura y funciones de aminoácidos péptidos y proteínas. En J.J. Hicks-Gómez (editor). Bioquímica. 2ª edición. MCGraw-Hill Interamericana. pp. 53-68.

Balogun, R.O., R.J. Jones y J.H.G. Holmes. 1998. Digestibility of some tropical browse species varying in tannin content. Anim. Feed Sci. Technol. 76: 77-83.

Bangs, P.D., Paul R. Krausman, P.R., Kunkel, K.E. and Parsons, Z.D. 2005. Habitat use by female desert bighorn sheep in the Fra Cristobal Mountains, New Mexico, USA. European J. of Wildlife Research. 51: 73-81.

Barahona, R., C.E. Lascano, R. Cochran, J. Morill, y E.C. Titgemeyer. 1997. Intake, digestion, and nitrogen utilization by sheep fed tropical legumes with contrasting tannin concentration and astringency. J. Anim. Sci. 75:1633-1640.

Barnes, T.G. 1988. Digestión dinamics in white-tailed deer. PhD. Thesis. Texas A&M University, Collage Station, p. 153.
Barnes, T.G., L.W. Varner, L.H. Blankenship, T.J. Fillinger, and S.C. Heineman. 1990. Macro and trace mineral content of selected south Texas deer forages. Journal of Range Management 43:220-223.
Bello J., Gallina S,. & Equihua M. 2001. Characterization and habitat preferences by whitetailed deer (Odocoileus virginianus) in Mexico with high drinking water availability. J. Range Manage. 54: 537-545.
Bello, J. 2001. Comportamiento del venado cola blanca texano en sitios con distintos manejo del agua en el noreste de México. Tesis Doctoral, Instituto de Ecología A. C., Ver.
Bento, M.H.L., T. Acamovic, H.P.S. Makkar. 2005. The influence of tannin, pectin and polyethylene glycol on attachment of 15N-labelled rumen microorganisms to cellulose. Anim. Feed Sci. Technol. 122:41-57
Bergman, E. N. 1990. Energy contributions of volatile fatty acids from the gastrointestinal
Bevans, O.J., Muller, C.G. y Watzka C.R. 2006. Vitamin K epoxide reductase complex subunit 1 (VKORC1): the key protein of the vitamin K cycle. Review. Antioxidants and Redox Signaling. 8: 347-53.
Biehl, J.C. y Lewis, t. 2003. Effects of forest cover on white-tailed deer migration in Northern Wisconsin Ohio Journal of Science.
Blümmel, M., Bullerdieck, P., 1997. The need to complement in vitro gas production measurements with residue determinations from in sacco degradabilities to improve the prediction of voluntary intake of hays. Anim. Sci. 64, 71-75.
Blümmel, M., Orskov, E.R., 1993.Comparison of in vitro gas production and nylon bag degradability of roughages in prediction of feed intake in cattle. Anim. Feed Sci. Technol. 40, 109-119.
Borba, A. E., Goncalves, P. M., Vouzela, C. F., Rego, O. A. y Borba, A, F. 2000. Effect of donor feeding in the use of alternative sources of inocula in the prevision of digestibility by the gas production method. Revista Portuguesa de Zootecnia. 1: 43-50.
Boroski BB & Mossman A. 1996. Distribution of mule deer in relation to water sources in Northern California. J. Wildl. Manage. 60: 770-776.
Boudet, A.M. 1998. A new view of lignification. Trens in Plant Science. 3:67-71
Bourgaud, F., A. Gravot, , S. Milesi, y E. Gontier. 2001. Production of plant secondary metabolites: a historical perspective. Plant Sci, 161:839-851.
Bowyer, R. T., y J. G. Kie. 2004. Effects of foraging activity on sexual segregation in mule deer. Journal of Mammalogy. 85:498-504.
Brady, P., L. Brady, P. Whetter, D. Ullrey, L. Fay. 1978. The effect of dietary selenium and vitamin E on biochemical parameters and survival of young

among white-tailed deer (Odocoileus virginianus) J. of Nutrition, 108: 1439-1448.
Brody, T. 1999. Nutritional Biochemistry. Academic Press. San Diego California, EUA. pp. 345-356.
Brown D.E. (1984) The effects of drought on whitetailed deer recruitment in the arid southwest. pp 7-12. In: P.R. Krausman y N.S. Smith (Ed.). Deer in the southwest: a workshop. Arizona Cooperative Wildlife Research Unit. University of Arizona. USA Davis E (1990) Deer management in the southern Texas plains. Texas Park and Wildlife Department. Federal Aid Reports Series. No. 27. Austin, Tex.
Brown, R.D. 1990. Nutrition y antler development: En: Horns, pronghorns and antlers. G.A. Bubenik y A.B. Bubenik (editores). Springer-Verlag, New York. pp. 426-441.
Brown, RD, 1994. The nutrients for survival and growth. In: Deer, D. Garlach, S. Atwater and J. Schnell, eds. Stackpole Books, Mechanicsburg, PA., EUA. pp. 203-207.
Brown, RD, 1994. The nutrients for survival and growth. In: Deer, D. Garlach, S. Atwater and J. Schnell, eds. Stackpole Books, Mechanicsburg, PA., EUA. pp. 203-207.
Bubenik, G.A., Sempere, A.J. y Hamr, J. 1987. Developing antler, a model for endocrine regulation of bone growth. Concentration gradient of T3, T4, and alkaline phosphatase in the antler, jugular, and the saphenous veins. Calcif Tissue Int. 41: 38:43.
Buchanan, B., Gruissem, W. Jones, R. 2000. Biochemistry and Molecular Biology of Plants. American society of plant physiology. John Wiley & Sons Ltd., Nueva York, EUA. pp. 245-266.
Bugalho, MN; Dove, H; Kelman, W; Wood, JT; Mayes, RW. 2004. Plant wax alkanes and alcohols as herbivore diet composition markers. Rangeland Ecology and Management 57 (3):259-268.
Burns, J.C., K.R. Pond, and D.S. Fisher. 1994. Measurement of forage intake. p. 494-526. In G.C. Fahey, Jr. et al. (ed.) Forage quality, evaluation, and utilization. ASA, CSSA, and SSSA. Madison, WI.
Buxton, D.R. and Fales, S.L. 1994. Plant environment and quality. In: G.C. Fahey Jr. Editor in Chief, National Conference on Forage Quality, Evaluation, and Utilization, University of Nebraska, Lincoln, NE. pp. 155-199.
Buxton, D.R., and J.R. Russell. 1988. Lignin constituents and cell wall digestibility of grass and legume stems. Crop Sci. 28:553.
Campa, H. III, Woodyard, D.K. y Hauler, J.B. 1984. Reliability of captive deer and cow in vitro digestion values in predicting wild deer digestion levels. J. Range Management, 37: 468-470.
Campbell, R.K. 2006. A Critical Review of Chromium Picolinate and Biotin. U.S. Pharmacist 31: 123-132.

Campbell, T. A. 1999. Antler development and nutritional influences of plant secondary compounds in mature white-tailed deer. Unpublished M.S. thesis, Texas A&M University-Kingsville, Kingsville.

Campbell, T.A. y Hewitt D.G. 2004. Mineral metabolism by white tailed deer fed diets of guajillo. The Southwestern Naturalist. 49: 367-375.

Carpita, N. y McCann, M. 2000. Biochemistry & Molecular Biology of Plants. American Society of Plant Physiologists. Rockville, Meriland. USA.

Cash, V.W. y Fulbright, T.E. 2005. Nutrient enrichment, tannins and thorns: effects on browsing of shrub seedlings. Journal of Wildlife Management 69:782-793. 2005

Castellanos R.A, Llamas L.l.G. y Shimada, S.A. 1990. Manual de técnicas de investigación en rumiología. Sistema de educación continua en producción animal en México, A. C. México, D. F. pp. 30-232.

Cheeke, P.R. 1998. Nartural toxicants in feeds, forages and poisonous plants. Segunda Edición. Interstate Publ., Inc., Danville, Illinois. Estados Unidos.

Chew, B.P. 1995. Antioxidant action of carotenoides. Jouranl of Nutrition, 119: 109-115.

Chiy, P.C. and C.J.C. Phillips. 1996. Sodium nutrition of dairy cows. In: Phillips, C.J.C. (ed.) Progress in Dairy Science. CAB International, Wallingford, pp. 29-44.

Church DC, Pond WG. 1996. Fundamentos de nutrición y alimentación de animales. Editorial Limusa, México, D. F. pp. 19-330.

Church, D.C. 1988. The Rumiant Animal Digestive Physiology and Nutrition. Prentice Hall, Englewood Cliffs, New Jersey, USA.

Clark, F.E. and Woodmansee, R.G. 1992. Nutrient cycling. In: Coupland, R.G. (ed.) Natural Grasslands: Introduction and Western Hemisphere, Elsevier, Amsterdam, pp. 137-146.

Clauss, M. and Lechner-Doll, M. 2001. Differences in selective reticulo-rumianl particle retention as a key factor in ruminant diversification, Oecology, 129: 321-327.

Clement, B. A., C. M. Goff, and T. D. A. Forbes. 1997. Toxic amines and alkaloids from Acacia berlandieri. Phytochemistry 46:249-254.

Coates, D.B. y Penning, P. 2000. Measuring animal performance. In: Field and Laboratory Methods for Grassland and Animal Production Research. Editado por L. `t Mannetje y R.M. Jones. CABI International. pp. 353-402.

Coleman, S.W., M. J. Williams, D.G. Riley, and C.C. Chase, Jr. 2004. The value of forage quality in winter feeding systems. Proc. STARS Forage/Beef Field Day. May 19, 2004. Brooksville, FL.

Colleman, S.W. y Henry, D.A. 2002. Nutritive of herbage. En: M. Freer y H. Dove (editores). Sheep Nutrition. CABI Publishing in association with CSIRO Publishing. Wallingford, Reino Unido. pp. 1-26.

Conradt, L. 1999. Social segregation is not a consequence of habitat segregation in red deer and feral Soay sheep. Animal Behaviour. 57:1151-1157.

Cooper, S. M., and T. F. Ginnett. 2000. Potential effects of supplemental feeding of deer on nest predation. Wildlife Society Bulletin 28:660-666.

Cooper, S. M., R. M. Cooper, M. K. Owens, and T. F. Ginnett. 2002. Effect of supplemental feeding on use of space and browse utilization by white-tailed deer. Page 31 in D. Forbes and G. Piccinni, editors. Land use for water and wildlife. Texas Agricultural Research and Extension Center, UREC-02-031, Uvalde, USA.

Correll D.S., M.C. Johnston, 1970, Manual of the Vascular Plants of Texas, Published by Texas Research Foundation, Volumen 6, pp. 1420

Cortés, C; Damasceno, JC; Pine, RC. 2005. Use of N-alkanes for estimations of botanical composition in samples with different proportions of Brachiaria brizantha and Arachis pintoi. Revista Brasileira de Zootecnia 34:1468-1474

COTECOCA. 1973. Comisión Técnica Consultiva para la Determinación Regional de los Coeficientes de Agostadero. SARH. México.

Cronjé, P. y Boomker. E.A. 2000. Ruminant Physiology: Digestion, Metabolism, Growth, and Reproduction. Wallingford, Oxfordshire, Reino Unido CABI Publishing.

Croteau, R. T.M. Kutchan, N.G. Lewis. 2000. Natural Products (Secondary Metabolites)". En: Buchanan, Gruissem, Jones (editores). Biochemistry and Molecular Biology of Plants. American Society of Plant Physiologists. Rockville, Maryland, Estados Unidos de América.

Cunnane, S.C. 2003. Problems with essential fatty acids: time for a new paradigm?. Progress in Lipid Research. 42: 544-568

Curtis, PS, and ME Richmond. 1992. Future challenges of suburban white-tailed deer management. Trans. N. Amer. Wildl. And Nat. Res. Conf. 57: 104-114.

Czerkawski, J. W. y K. J. Cheng. 1988. Compartimentation in the rumen. En: P. N. Hobson (Ed.). The Rumen Microbial Ecosystem . Elsevier Applied Science, Nueva York. pp. 361-385.

D'Mello, J.P.F. 2000. Anti-nutritional factors and mycotoxins. En: J.P.F. D'Mello (editor). Farm Animal Metabolism and Nutrition. CABI Publishing. Wallingford, Reino Unido. pp. 393-403.

Davis, G.K. and W. Mertz. 1987. Copper. In: Mertz, W. (ed.) Trace Elements in Human and animal Nutrition, 5th edn, Vol. 1 Academic Press, San Diego, pp. 301-364.

DeBolt, D.C. 1980. Mulielement emisión spectroscopy analisis of plant tissue using DC argon plasma source, J. Assoc. Off, Agric. Chem. 63: 802-805.

Decandia, M., M. Sitziaa, A. Cabiddua, D. Kababyab y G. Molleag. 2000. The use of polyethylene glycol to reduce the anti-nutritional effects of tannins in goats fed woody species. Small Rumin. Res. 38:157-164.

Dehority, ,B.A. 1994. Rumen ciliate protozoa of the blue duiker, with observations on morphological variation lines within the species Entodinium dubardi. J. Eukaryot Microbiology, 41: 103-111.

Dehority, B.A. 1993. Microbial ecology of cell wall fermentation. p. 425-453. In H. gG. Jung, D.R. Buxton, R.D. Hatfield, and J. Ralph (ed.) Forage cell wall structure and digestibility. ASA-CSSA-SSSA, Madison, WI.

Devendra, C. 1990. The use of shrubs and tree fodders by ruminants. En: Shrubs and tree fodders for farm animals. C. Devendra, editor, Proceedings of a Workshop in Denpasar, Indonesia, IDRC 276e, Ottawa, Canada,. pp. 88-96.

Dierenfeld, E.S. 1989. Vitamin E deficiency in zoo reptiles, birds, and ungulates. J. of Zoology Wildlife, 20: 3-11.

Dijkstra, J., Forbes, J.M. y France, J. 2005. Quantitative Aspects of Ruminant Digestion and Metabolism, 2nd edition. Wallingford, Oxfordshire, Reino Unido. CABI Publishing.

Dinkines, W.C., Lochmiller, R.L., Bartus, W.S., y Qualis, C.W. Jr. 1991. Using condition indicators to evaluate habitat quality for white-tailed deer. Procedings of the Annual Conference of the Southeastern Association of Fish and Wildlife Agencies 45:19-29.

Domínguez-Bello, M.G. 1996. Detoxification in the rumen. Ann. Zootech., 45, suppl.: 323-327.

Doreau, M. and Y. Chilliard. 1997. Digestion and metabolism of dietary fat in farm animals. British J. of Nutrition 78 (suplemento 1), S15-S35.

Dove, H. 1996. The ruminant, the rumen and the pasture resource: Nutrient interactions in the grazing animal. En: J. Hodgson and A.W. Illius (editores) The Ecology and Management of Grazing Systems. CAB Intl., Wallingford, OX, Reino Unido, pp. 219-246.

Dove, H; Mayes, RW. 1991. The use of wax alkanes as marker substances in studies of the nutrition of herbivores: a review. Australian Journal of Agricultural Research 42:913-952.

Duncan, A.J. y Poppi, D.P. 2008. Nutritional ecology of grazing and browsing ruminants. En: I.J. Gordon y H.H.T. Prins (Editores). The Ecology of Browsing and Grazing. Springer. Verlang, Berlin Heilderbeg. pp. 89-111.

Dunkan, A.J., Ginane, C. Gordon, I.J. y Orskov, E.R. 2003. Why herbivores select mixec diets?. En: L. 't Mannetje, L. Ramírez-Avilés, C. Sandoval-Castro, J.C. Ku-Vera (editors). Matching Herbivore Nutrition to Ecosystems Biodiversity. Proceedings of VI International Symposium on the Nutrition of Herbivores. Mérida, Yucatán, México.

Durand, M. and S. Komisarczuk. 1988. Influence of major minerals in rumen microbiota. J. of Nutrition., 118: 249-260.

Engels, F.M. 1989. Some properties of cell wall layers determining ruminant digestion. In: A. Chesson and E.R. Orskov (Ed.) Physico-Chemical

Characterization of Plant Residues for Industrial and Feed Use. p 80. Applied Science Publishers, London.
Etzenhouser, M.J., Owens, M.K., Spalinger, D.E. y Murden, S.B.. 1998. Foraging behavior of browsing ruminants in a heterogeneous landscape. Landscape Ecology 13:55-64.
Everitt, J.H., and C.L. Gonzales. 1981. Seasonal nutrient content in food plants of white-tailed deer on the south Texas Plains. Journal of Range Management 34:506-510.
Everitt, J.H., D. Lynn and R.I. Lonard. Trees, Shrubs and Cacti of south Texas. 2002. Revised Edition. Texas Tech University Press, p. 99.
Feist, M.S. 1998. Evaluation of development of specialized livestock diets in Saskatchewan. MSc. Thesis. Universitry of Saskatchewan.
Fell, D. 1997. Understanding the control of metabolism. Portland Press. Londres, pp. 134-145.
Forbes, J.M. 2007. Voluntary Food Intake and Diet Selection in Farm Animals. 2a Edición. pp. 11-37.
Forbes, J.M. y Provenza, F.D. 2000. Integration of Learning and Metabolic Signals into a Theory of Dietary Choice and Food Intake. En: J.A. Cronjé (editor). Ruminant Physiology Digestion, Metabolism, Growth and Reproduction. CABI Publishing. Wallingford, Reino Unido. pp. 3-20.
Forbes, T.D.A., and S.W. Coleman. 1993. Forage intake and ingestive behavior of cattle grazing old world bluestems. Agron. J. 85:808-816.
Foroughbachkch R., R.G. Ramírez, L.A. Háuad y J. G. Moya-Rodríguez. 1998 .Dinámica estacional de la degradabilidad ruminal de nutrientes de 10 arbustivas nativas del noreste de México. International Journal of Experimental Botany, YTON, 63:179-186.
Forsberg, C.W., Forano, E. y Chesson, A. 2000. Microbial Adherence to the Plant Cell Wall and Enzymatic Hydrolisis. En: P.B. Cronjé (editor). Ruminant Physiology, Digestion, Metabolism, Growth and Reproduction. CABI, Publishing, Wallingford, Reino Unido. pp. 79-98.
Fowler, M.E. 1983. Plant poisoning in free-living wild ruminants: a review. J. Wildl. Dis. 19: 34-43.
France, J., Dijkstra, J., Dhanoa, M.S., Lopez, S., Bannik, A., 2000. Estimating the extent of degradation of ruminant feeds from a description of their gas production profiles observed in vitro: derivation of models and other mathematical considerations. Br. J. Nutr. 83, 143-150.
Freeland, W.J. Calcott P.H. y Geiss, D.P. 1985. Allelochemicals, minerals and herbivore population size. Biochemical Systematics and Ecology. 13: 195-204.
Gallina, S. 1993. White-tailed deer and cattle diets at La Michilia Durango, México, J. of Range Management, 46: 486-492.

Gallina, S. y Bello, J. 2010. El Gasto Energético del Venado Cola Blanca (Odocoileus virginianus texanus) en Relación a la Precipitación en una Zona Semiárida de México. Therya. 1: 9-22

García Herrera, N. 1994. Valor Nutritivo y Digestibilidad in situ de la Materia Seca y Proteína Cruda del Forraje de 13 Hierbas Nativas de Nuevo León, Colectadas en Otoño. Tesis Profesional. Facultad de Medicina Veterinaria y Zootecnia, Universidad Autónoma de Nuevo León.

García Herrera, N. 1994. Valor Nutritivo y Digestibilidad in situ de la Materia Seca y Proteína Cruda del Forraje de 13 Hierbas Nativas de Nuevo León, Colectadas en Otoño. Tesis Profesional. Facultad de Medicina Veterinaria y Zootecnia, Universidad Autónoma de Nuevo León.

García Martínez, P.M. 1995. Valor nutritivo y digestibilidad in situ de la materia seca y proteína cruda de siete hierbas nativas de noreste de México. Tesis Profesional, Facultad de Medicina Veterinaria y Zootecnia, Universidad Autónoma de Nuevo León.

Garin, I., A. Aldezabal, R. García-González y J. R. Aihartza. 2001. Composición y calidad de la dieta del ciervo (Cervus elaphus L.) en el norte de la península ibérica. Animal Biodiversity and Conservation. 24:53-63.

Gessman, J.A. 1987. Energetics. En: Raptor Management Techniques Manual. B.A. Giron Pendelton, B.A, Millsap, K.W. Clire y D.A. Birds (editores). Natl. Wildl. Fed., Washington, DC. pp. 289-320.

Gibbons, B.J., Roach, P.J., and Hurley, T.D. 2002. Crystal structure of the autocatalytic initiator of glycogen synthesis, glycogenin. Journal of Molecular Biology. 319:463-477.

Goff JP. 2000. Pathophysiology of calcium and phosphorus disorders. Vet Clin North Am Food Pract 16, 319-337.

González-Gómez, J.C., A. Ayala-Burgos y D. Gutiérrez-Vázquez. 2006. Determinación de fenoles totales y taninos condensados en especies arbóreas con potencial forrajero de la Región de Tierra Caliente Michoacán, México. Livestock Research for Rural Development. Volume 18, Article No. 152. Obtenido en Febrero 10, 2010, de:

Grace, N.D. and R.G. Clark. 1991. Trace element requirements, diagnosis and prevention of deficiencies in sheep and cattle. In: Tsuda, T., Sasaki, Y. and Kawashima, R. (eds.) Physiological aspects of Digestion and Metabolism in Ruminants. Academic Press, San Diedo, pp. 321-346.

Grant R J and Mertens D R 1992a Development of buffer systems for pH control and evaluation of pH effects on fiber digestion in vitro. Journal of Dairy Science. 75: 1581-1587.

Grant R J and Mertens D R 1992b Impact of in vitro fermentation techniques upon kinetics of fiber digestion. Journal of Dairy Science. 75: 1263-1272.

Grasman, B.T. and E.C. Hellgren. 1993. Phosphorous nutrition in white-tailed deer: nutrient balance, physiological responses, and antler growth. Ecology 74:2279-2296.

Guerrero Cervantes, M. 2009. Valor nutricional de forrajes nativos del noreste de México. Tesis de Doctorado. Facultad de ciencias Biológicas, Universidad Autónoma de Nuevo León. San N icolas de los Garza, Nuevo León, México.

Guerrero-Cervantes, M., R.G. Ramírez, M.A. Cerrillo-Soto, R. Montoya-Escalante, G. Nevárez-Carrasco y A.S. Juárez-Reyes. 2009. Dry matter digestion of native forages consumed by range goats in North Mexico. J. Anim. Vet. Adv. 8:408-412.

Guimaraes-Beelen, P.M., T.T. Berchielli, R. Beelen y A.N. Medeiros. 2006. Influence of condensed tannins from Brazilian semi-arid legumes on ruminal degradability, microbial colonization and ruminal enzymatic activity in Saanen goats. Small Rumin. Res. 61:35-44.

Haenlein, G.F.W. and Ramírez, R.G. 2007. Potential mineral deficiencies on arid rangelands from small ruminants with special reference to Mexico. Small Rumin. Res. 68: 35:41.

Halls, LK. 1984. Whit-tailed deer: Ecology and Management. Stackpole Books, Harrissburg, PA. EUA.

Hatfield R.D., J. Ralph, J. Grabber, and H.J. Jung. 1993. Structural characterization of isolated corn lignins. p. A-319. In Abstr. Keystone Symposia, the extracellular matrix of plants: Molecular, cellular and developmental biology. Santa Fe, NM, 9-15 Jan.

Hatfield, R.D., H.G. Jung, J. Ralph, D.R. Buxton, and P.J. Weimer. 1994. A comparison of the insoluble residues produced by the Klason lignin and acid detergent lignin procedures. J. Sci. Food Agric. 65:51.

Haufler, J.B. y Servello, A. 1996. Técnicas for wildlife nutritional analyses. En: Research and Management Techniques for Wildlife and habitats. T.A. Bookhout (editor). The Wildlife Society. pp. 307-323.

Heaney, D.P. 1970. Voluntary intake as a component of an index to forage quality. p. C1-C10. In R.F. Barnes, D.C. Clanton, C.H. Gordon, T.J. Klopfenstein, and D.R. Waldo (Ed.) Proc. Natl. Conf. Forage Qual. Eval. Util., Lincoln, NE.

Hellgren, E. C., and W. J. Pitts. 1997. Sodium economy in white-tailed deer (Odocoileus virginianus). Physiological Zoology 70:547-555.

Hernández G, R. 1997. Análisis estructural e importancia económica de Helieta parvifolia Gray (Benth), en dos zonas ecológicas en el estado de Nuevo León. Tesis (profesional) Facultad de Ciencias Biológicas, UANL. San Nicolás de los Garza , Nuevo León, México., pp 6-16, 35.

Hernández Romero, A.L. 1995. Digestión ruminal de la proteína cruda y pared celular de 9 hierbas del estado de Nuevo León, colectadas en verano. Tesis

Profesional, Facultad de Medicina Veterinaria y Zootecnia, Universidad Autónoma de Nuevo León.

Hervert, J.J. y Krausman, P.R. 1986. Desert Mule Deer Use of Water Developments in Arizona The Journal of Wildlife Management, 50: 670-676

Hespell, R.B. 1988. Microbial digestion of hemicelluloses in the rumen. Microbiol. Sci. 5: 362-365.

Hicks-Gómez, J.J. 2007. El agua: propiedades y capacidad disolvente de las biomoléculas. En J.J. Hicks-Gómez (editor). Bioquímica. 2ª edición. MCGraw-Hill Interamericana. pp. 36-52.

Hidayat, C., Hillman, K., Newbold, C. J. y Stewart, C. S. 1993. The contributions of bacteria and protozoa to ruminal forage fermentation in vitro, as determined by microbial gas production. Animal Feed Science and Technology. 42: 193-208.

Hobson, P.N. y Stewart, C.S. 1997. The Rumen Microbial Ecosystem, 2nd edition. New York: Springer. of sward characteristics on diet selection and herbage by the grazing animal. p. 153-166. In J.B. Hacker (ed.) Nutritional limits to animal production from pastures. CAB, Slough, UK.

Hofman, R.R. 1989. Evolutionary steps of ecophysiological adaptation and diversification of ruminants: a comparative viewnof their digestive system. Oecologia, 78: 443-457.

Hofmann, R.R. 1988. Anatomy of the Gastro-Intestinal Tract. En: The ruminant animal, Digestive Physiology and Nutrition. D.C. Church (editor). Waveland Press, Inc. Illinois, Estados Unidos. pp. 14-43.

Holechek, J.L., Pieper, R.D. y Herbel, C.H. 2001. Range Management: Principles and Practices. Fourth edition. Prentice Hall, Upper Saddle River, New Jersey, USA.

Holechek, J.L., Vavra, M. y Pieper, R.D. 1982. Botanical composition determination of range herbivore diets: a review. J. Range Management 35: 309-315.

Holloway, J. W., R. E. Estell, II, and W. T. Butts, Jr. 1981. Relationship between fecal components and forage consumption and digestibility. J. Anim. Sci. 52:836-848.

Holmes, J.H.G., M.C. Franklin, and L.J. Lambourne. 1966. The effects of season, supplementation, and pelleting on intake and utilization of some sub-tropical pastures. Proc. Aust. Soc. Anim. Prod. 6:354-363.

Horst, R.L. 1986. Regulation of calcium and phosphorous homeostasis in the dairy cow. J. Dairy Science, 69:604-616.

Horton, H.R., Moran, L.A., Ochs, R.S., Rawn, J.D., Scrimgeour, K.G. 1995. Bioquímica. México, D.F: Prentice-Hall Hispanoamericana.

Hoste, H., F. Jackson, S. Athanasiadou, S.M. Thamsborg y S.O. Hoskin. 2006. The effects of tannin-rich plants on parasitic nematodes in ruminants. Trends Parasitol. 22:253-261.

Iason, G.R., J. Hodgson y T.N. Barry. 1995. Variation in condensed tannin concentration of a temperate grass (Holcus lanatus) in relation to season and reproductive development. J. Chem. Ecol., 21:1103-1112.

Ibarra, F.A., M.H. Martín, H. Miranda, and J.L. Luna. 1998. Seeding of Forage Brush Species for the Restoration of Deteriorated Rangelands in the Sonoran Desert. Soc. Range Manage. Meeting. Guadalajara, Jal. México.p.63.

Iiyama, K., T.B.T. Lam, P.J. Meilke, D.I. Rhodes, and B.A. Stone. 1993. Cell Wall biosynthesis and its regulation. In: H.G. Jung, D.R. Buxton, R.D. Hatfield, and J. Ralph (Ed.) Forage Cell Wall Structure and Digestibility. p 621. ASA-CSSA-SSSA, Madison, WI.

Illus, A. W., y J. I. Gordon. 1999. The physiological ecology of mammalian herbivore. In: H.G. Jung y G. C. L Fahey. Jr., (eds.). Nutritional ecology of herbivores. American Society of Animal Science, Savoy, Illinois. pp. 71-96.

Jackson, F.S., T.N. Barry, C. Lascano y B. Palmer. 1996. The extractable and bound condensed tannin content of leaves from tropical tree, shrub and forage legume. J. Sci. Food Agric., 71:103-110.

Jarrige, R., C. Demarquilly, J. P. Dulphy, A. Hoden, J. Robelin, C. Beranbger, Y. Geay, M. Journet, C. Malterre, D. Micol, and M. Petit. 1986. The INRA "fill unit" system for predicting the voluntary intake of forage-based diets in ruminants: A review. J. Anim. Sci. 63:1737-1758.

Jean-Blain, C. 1998. Aspectes nutritionnels et toxicologiques des tannins. Rev. Méd. Vét. 149;911-920.

Jenks, JA, DM Leslie, Jr. y RL Lochmier. 1990. Food habits and nutritional condition of white-tailed deer and cattle. Final Rep. PR Project W-142-R, Okla. Dep. Wild. Conserv., Oklahoma City, OK. pp. 91-96.

Johns, P.E., E.G. Cothran, M.H. Smith, and R.K. Chesser. 1982. Fat levels in male white-tailed deer during the breeding season. Proceedings of the Annual Conference of Southeastern Association of Fish and Wildlife Agencies 36:454-462.

Johnson, L.E. 2007. Vitamin D: The Merck Manual of Diagnosis and Therapy. Last full review.

Jones, P.D., Demarais, S., Strickland, B.K. y Edwards, S.L. 2008. Soil region effects on white-tailed deer forage protein content. Southeastern Naturalist 7: 595-606.

Jones, R.M., Bishop, H.G., Clem, R.L., Conway, M.J. y Cook, B.G. 2000. Measurements of nutritive value of a range of tropical legumes and their use in legume evaluation. Tropical Grasslands, 34: 78-90.

Jones, S.D., L.K. Wipff and P.M. Montgomery. 1997. Vascular Plants of Texas: A comprehensive checklist including synonymy, bibliography, and index. University of Texas, Press. Austin.

Jouany, J.P. 1991. Defaunation of the rumen. En: J.P. Jouany (editor). Rumen Microbial Metabolism and Ruminant Digestion. INRA, Francia, pp. 239-262.

Judd, W.S., Campbell, C.S., Kellogg, E.A., Stevens, P.F. y Donoghue, M.J. 2002. Plant Systematics: A Phylogenetic Approach. Segunda Edición. Sinauer Axxoc, Estados Unidos de América.

Jung, H.G., and D.A. Deetz. 1993. Cell wall lignification and degradability. In: H.G. Jung, D.R. Buxton, R.D. Hatfield, and J. Ralph (Ed.) Forage Cell Wall Structure and Digestibility. p 315. ASA-CSSA-SSSA, Madison, WI.

Jung, H.G., and K.P. Vogel. 1992. Lignification of switchgrass (Panicum virgatum L.) and big bluesteam (Andropogon gerardii Vitman) plant parts during maturation and its effect on fibre degradability. J. Sci. Food Agric. 59:169.

Jung, H.G., and M.D. Casler. 1991. Relationship of lignin and esterified phenolics to fermentation of smooth bromegrass fibre. Anim. Feed Sci. Technol. 32:63.

Jung, H.G., and M.S. Allen. 1995. Characteristics of plants of cell walls affecting intake and digestibility of forages by ruminants. J. Anim. Sci. 73:2774-2790.

Jung, H.G., R.R. Smith, and C.S. Endres. 1994. Cell wall composition and degradability of stem tissue from lucerne divergently selected for lignin and in vitro dry matter disappearance. Grass Forage Sci. 49:295.

Kamalak, A., O. Canbolat, Y. Gurbuz, O. Ozay, C.O. Ozkan y M. Sakarya. 2004. Chemical composition and in vitro gas production characteristics of several tannin containing tree leaves. Livestock Research for Rural Development, Vol. 16, Art.#44 . ht tp://www.cipav.org.co/lrrd/lrrd16/6/kama16044.htm.

Kerley, M.S., G.C. Fahey, Jr., J.M. Gould, and E.L. Iannotti. 1992. Effects of lignification, cellulose crystallinity and enzyme accessible space on the digestibility of plant cell wall carbohydrates by the ruminant. Food Microstructure 7: 59-65.

Kessler, J.J. 1990. Atriplex forage as a dry season supplementation feed for sheep in the montane plains of the Yemen Arab Republic, Journal of Arid Enviroments. 19:225-234.

Koch, A. 2003. Bacterial wall as target for attack: past, present, and future research. Clinical Microbiology Review. 16: 673 - 87.

Krishnamoorthy, U., Soller, H., Steingass, H.H., Menke, K.H., 1991. A comparative study on rumen fermentation of energy supplements in vitro. J. Anim. Physiol. Anim. Nutr. 65, 28-35.

Kroll, J. C. 1992. A practical guide to producing and harvesting white-tailed deer. Institute of White tailed deer management and research center for applied studies in forestry. Stephen F. Austin State University. Austin, Texas.

Kucera, E.T. 1997. Fecal indicators, diet, and population parameters in mule deer. Journal of Wildlife Management 61: 550-560.

Kumara-Mahipala, M.B.P., G.L. Krebs, P. McCafferty y L.H.P. Gunaratne. 2009. Chemical composition, biological effects of tannin and in vitro nutritive value of selected browse species grown in the West Australian Mediterranean environment. Anim. Feed Sci. Technol. 153:203-215.

Langer, P. y Snipes, R.L. 1991. Adaptations of Gut Structure to Function in Herbivores. En: T. Tsuda, Y. Sasaki y R. Kawashima (editors). Physiological Aspects of Digestion and Metabolism in Ruminants. Academic Press, Inc. San Diego, Estados Unidos. pp. 349-384.

Lason, G.R. and Van Wieren, S.E. 1998. Adaptations of mammalian herbivores to low quality forage. In: Olff H., Brown, V.K., Drent, R.H. (editores) Herbivores, plants and predators. Blackwell, Oxford, pp. 337-369.

Lautier, J.K., Dailey, T.V. y Brown, R.D. 1988. Effect of Water Restriction on Feed Intake of White-Tailed Deer. Journal of Wildlife Management. 52: 602-606

Le Houerou, H.N., 2000. Utilization of Fodder Trees and Shrubs in the Arid and Semiarid Zones of West Asia and North Africa. Arid Soil Res.Rehab. 14:101-135.

Lehninger, A. L. 1976. Curso breve de Bioquímica. Omega, Barcelona, pp. 447.

Lehninger, A.L., Cox, M.M. y Nelson, D.L. 2000. Principles of Biochemistry. W.H. Freeman & Co Worth Publishers. New York, EUA.

Lehninger, A.L., Cox, M.M. y Nelson, D.L. 2000. Principles of Biochemistry. W.H. Freeman & Co Worth Publishers. New York, EUA.

Leite, E.R., and J.W. Stuth. 1990. Value of multiple fecal indices for predicting diet quality and intake of steers. J. Range Manage. 43:139-143.

Leng, R.A., S.H. Bird, A. Klieve, B.S. Choo, F.M. Ball, G. Asefa, P. Brumby, V.D. Mudgal, U.B. Chaudry, N. Suharyano. 1992. The potential for tree forage supplements to manipulate rumen protozoa to enhance protein to energy ratios in ruminants fed on poor quality forages. In: Legume Trees and Other Fodder Trees as Protein Sources for Livestock, FOA Animal Production and Health Paper 102. pp. 177-191 Rome: FAO.

Leng, RA. 1997. Tree Foliage in Ruminant Nutrition. FAO Animal Production and Health. Paper 139, pp. 69-79.

Leslie, D.M., Jr., y E. E. Starkey 1985. Fecal indices to dietary quality: a reply. The Journal of Wildlife Management. 51: 321-325.

Lincoln, T. y E. Zeiger. 2006. Secondary Metabolites and Plant Defense. Plant Physiology, Fourth Edition. Sinauer Associates, Inc. EUA.

Lippke, H. 1980. Forage characteristics related to intake, digestibility and gain by ruminants. J. Anim. Sci. 50:952-961.

Litwack, G. 2007. Vitamin A (Vitamins and Hormones). Elsevier Academic Press, San Diego, CA, EUA. pp. 412.

Lobley, G.E. Amino acid and protein metabolism in the whole body and individual tissues of the ruminant. 2000. En: J.M. Asplund (Editor). Principles of Protein Nutrition of Ruminants. CRC Press, Boca Raton, FL, EUA, pp. 79-97.

López, S., Carro, M. D., González, J. S. y Ovejero, F.J. 1998. Comparison of different in vitro and in situ methods to estimate the extent and rate of degradation of hays in the rumen. Animal Feed Science and Technology. 73: 99-113.

Loredo, M.A., G.D. Simpson, D.J. Minson, and C.G. Orpin. 1991. the potential for using n-alkanes in tropical forages as a marker for the determination of dry matter intake by grazing ruminants. J. Agric. Sci., Camb., 117:355-361.

Lovegrove, B. G. 2000. The zoogeography of mammalian basal metabolic rate. American Naturalist 156:201-219.

Lowry, B.J., J.R. Petherman and B. Tangengjaja. 1992. Plants fed to village ruminant in Indonesia. ACIAR Technical Report No. 22, Canberra, p. 60.

Mackie, R.I., Aminov, R.I., White, B.A. y McSwenney, C.S. 2000. Molecular Ecology and Diversity in Gut Microbial Ecosystems. En: P.B. Cronjé (editor). Ruminant Physiology, Digestion, Metabolism, Growth and Reproduction. CABI, Publishing, Wallingford, Reino Unido. pp. 61-78.

Mackie, R.I., C.S. McSweeney, A.V. Klieve. 2002. Microbial Ecology of the Ovine Rumen. En Freer, M., y H. Dove 2002 Sheep nutrition. Cabi publishing, Wallinford, Reino Unido. p. 385.

Macoon, B., L. E. Sollenberger, J. E. Moore, C. R. Staples, J. H. Fike, and K. M. Portier. 2003. Comparison of three techniques for estimating the forage intake of lactating dairy cows on pasture. J. Anim. Sci. 81:2357-2366.

Maghini, M.T. & Smith, N.S. (1990) Water use and diurnal seasonal ranges of Coues white-tailed deer. pp. 21-34. In: Krausman, P.R. & N.S. Smith Editors. Deer in the southwest: A workshop. Arizona Cooperative Wildlife Research Unit. University of Arizona Tucson.

Main, M. B., F. L. Werckely y V. V. Bleich. 1996. Sexual segregation in ungulates: New direction for research. Journal of Mammalogy. 77:449-461.

Makkar, H. 2001. Recent advances in in vitro gas method for evaluation of nutritional quality of feed resources. http://www.fao.org/DOCREP/ARTICLE/AGRIPPA/570_EN_toc.htm

Makkar, H.P.S. 2006. Chemical and biological assays for quantification of major plant secondary metabolites. BSAS Publication 34. The assessment of intake, digestibility and the roles of secondary compounds. Edited by C.A. Sandoval-Castro, F.D.DeB.D. Hovell, J.F.J. Torres-Acosta and A. Ayala-Burgos. Nottingham University Press. pp. 235-249.

Makkar, H.P.S., 2003. Effects and fate of tannins in ruminant animals, adaptation to tannins strategies to overcome detrimental effects of feeding tannin-rich feeds. Small Rumin. Res. 49:241-256.

Mandujano, S. y S. Gallina. 1995. Disponibilidad de agua para el venado cola blanca en un bosque tropical caducifolio de México. Vida Silvestre Neotropical, 4:107-118.

Masters, R, TG Bidwell and M Shaw, 1995. Ecology and Management of deer in Oklahoma. F-9009. Oklahoma Cooperative Extension Service. Division of Agricultural Siences and Natural Resources, Oklahoma State University, pp 1-9.

Matthews, J.C. 2000. Amino acid and peptide transport systems. En: J.P.F. D'Mello. Farm Animal Metabolism and Nutrition. CABI International. Wallingford, Reino Unido. pp. 3-24.

Mautz, W. W., J. Kanter y P. J. Pekins. 1992. Seasonal metabolic rhythms of captive female white-tailed deer: a re-examination. Journal of Wildlife Management 56:656-661.

Mayes, RW; Dove, H. 2006. The use of N-alkanes and other plantwax compounds as markers for studying the feeding and nutrition of large mammalian herbivores. In Sandoval-Castro, CA; DeB, H; Torres Acosta JF; Ayala, A. (Eds.). Herbivores: The assessment of intake, digestibility and the roles of secondary compounds. Nottingham, UK, Nottingham Univ. British Society of Animal Science Publication No. 34. p. 153-182.

Maynard, L. A., Loosli, J. K., Hints, H. F. and R. G. Warner. 1992. Nutrition Animal 4ta. Edición en español ed. McGraw-Hill, México.

McCullough, DR, Hirth, DH and Newhouse, SJ. 1989. Resource partitioning between sexes in white-tailed deer. J. Wildlife Management. 5: 277-282.

McDonald, I. 1981. A revised model for estimation of protein degradability in the rumen. J Agric Sci Camb 96:251-252.

McDowell, L.R. 1998. Reevaluation of the metabolic essentiality of the vitamins. Proceedings: New Technologies for the Production of "Next Generation" Feeds and additives. The 8th World Conference on animal Production, Seoul National University, Seoul Korea, pp. 54-77.

McDowell, R.L., J.H. Conrad, F.G. Hembry, L.X. Rojas, G. Valle y J. Velásquez. 1997. Minerales para rumiantes en pastoreo en regiones tropicales. 2a. edición. Departamento de Zootecnia, Universidad de Florida, Gainesville, Florida, EUA, p. 10.

McSweeney, C.S., H.P.S. Makkar y J.D. Reed. 2003. Modification of rumen fermentation to reduce adverse effects of phytochemicals. In Matching herbivore nutrition to ecosystems biodiversity. L. t'Mannetje, L. Ramírez-Avilés, C. Sandoval-Castro, J.C. Ku-Vera (Eds.). VI International Symposium on the Nutrition of Herbivores. pp. 239-268.

Merchen N.R. y Bourquin, L.D. 1994. Processes of digestion and factors influencing digestion of forage based diets by ruminants. En G.C. Fahey, Jr. (editor). Forage Quality, Evaluation, and Utilization. University of Nebraska, Lincoln, NE. Estados Unidos. Pp. 564-612.

Merchen N.R. y Bourquin, L.D. 1994. Processes of digestion and factors influencing digestion of forage based diets by ruminants. En G.C. Fahey, Jr.

(editor). Forage Quality, Evaluation, and Utilization. University of Nebraska, Lincoln, NE. Estados Unidos. Pp. 564-612.

Merchen, N.R. 1988. Digestion, Absorption and Excretion in Ruminants. En: The ruminant animal, Digestive Physiology and Nutrition. D.C. Church (editor). Waveland Press, Inc. Illinois, Estados Unidos. pp. 173-201.

Merchen, N.R. 1988. Digestion, Absorption and Excretion in Ruminants. En: The ruminant animal, Digestive Physiology and Nutrition. D.C. Church (editor). Waveland Press, Inc. Illinois, Estados Unidos. pp. 173-201.

Merla, R.G. 1990. Nuevo León, Geografía Regional. Universidad Autónoma de Nuevo león. Centro de Información de Historia Regional. Serie Biblioteca de Nuevo León/9.

Mersamann, H.J. Lípids. En: En: W.G. Pond y A.W. Bell (Editores). Encyclopedia of Animal Science. Marcel Dekker, NY, EUA. pp. 578-581.

Mertens, D. R. 1993. Kinetics of cell wall digestion and passage in ruminants. p. 535-570. En H. G. Jung, D. R. Buxton, R. D. Hatfield, and J. Ralph (ed.) Forage cell wall structure and digestibility. ASA-CSSA-SSSA, Madison, WI.

Mertens, D.R. 1993. Kinetics of cell wall digestion and passage in rumiants. p. 535-570. In H. G. Jung, D.R. Buxton, R.D. Hatfield, and J. Ralph (ed.). Forage cell wall structure and digestibility. ASA-CSSA-SSSA, Madison, WI.

Meyer, H.H.D., Rowell, A., Streich, W.J., Stoffel, B. and Hofmann, R.R. 1998. Accumulation of polyunsaturated fatty acids by concentrate selecting ruminants. Comp. Biochemistry Physiology, 120A: 263-268.

Michels, T.R., "The Whitetail Addicts Manual", Creative Publishing 2007, ISBN 978-1-58923-344-7

Miller, E.R., X. Lei, and D.E. Ullrey. 1991. Trace elements in animal nutrition. In: Mortvedt, J.J. (ed.) Micronutrients in Agriculture, 2nd edn. Soil Science Society of America, Madison, pp. 596-562.

Miller, K.V., R.L. Marchinton, J.R. Beckwith and P.B. Bush. 1985. Variations in density and chemical composition of white-tailed deer antlers. J. Mammal. 66: 693-701.

Min, B.R., G.T. Attwood, K. Reilly, W. Sun, J.S. Peters, T.N. Barry y W.C. McNabb. 2002. Lotus corniculatus condensed tannins decrease in vivo populations of proteolitic bacteria and affect nitrogen metabolism in the rumen of sheep. Can. J. Microb. 48: 911-921.

Minson, D.J. 1990. Forage in ruminant nutrition. Academic Press, New York, USA.

Moen, A. 1978. Seasonal changes in heart rates, activity, metabolism, and forage intake of white-tailed deer. Journal of Wildlife Management 42: 715-738.

Molina, D.O., A. N. Pell, y D.E. Hogue. 1999. Effects of ruminal inoculations with tannin-tolerant bacteria on fiber and nitrogen digestibility of lambs fed high-condensed tannin diet. Anim. Feed Sci. Technol. 81:69-75.

Monties, B. 1991. Plant cell as fibrous lignocellulosic composites: Relations with lignin structure and function. Anim. Feed Sci. Technol. 32:159-175.

Moore J.E., and D.J. Undersander. 2002. A proposal for replacing relative feed value with an alternative: relative forage quality. Proc. American Forage and Grassland Council 11:171-175.

Moore, J.E., J.C. Burns, and D.S. Fisher. 1996. Multiple regression equations for predicting relative feed value of grass hays. p. 135-139. In M.J. Williams (ed.) Proc. American Forage and Grassld. Council, 13-15 Jun 1996, Vancouver, BC., American Forage and Grassld. Council, Georgetown, TX.

Moore, J.H. and W.W. Christie. 1984. Digestion absorption and transport of fats in ruminant animals. In: Wiseman, J. (ed.) Fats in Animal Nutrition. Butterworths, Londres, pp. 123-149.

Moore, K and R. García, 2001). Role, impacts, and management of plant lignin. In: Antiquality Factors in Rangeland and Pastureland Forages. USDA, Natural Resources Conservation Services Grazing Lands Technology Institute, University of Idaho, Idaho, USA, pp. 23-27.

Morrison, I.M. 1980. Changes in the lignin and hemicellulose concentrations of ten varieties of temperate grasses with increasing maturity. Grass Forage Sci. 35:287-293.

Moya-Rodríguez, J. G., R.G. Ramírez and R. Foroughbakhch. 2002. Seasonal changes in cell wall digestion of eight browse species from northeastern Mexico. Revista Electrónica: Livestock Research for Rural Development 14:(1) 1-7.

Moya-Rodríguez, J.G., R.G. Ramírez, R. Foroughbackhch, L.A. Aguad y H. González-Rodríguez. 2002. Variación estacional de minerales en las hojas de ocho especies arbustivas. Ciencia UANL. 5: 59-65.

Moya-Rodríguez, R.G. Ramírez and Foroughbackhch, R. 2003. Nutritional value and effective degradability of crude protein in browse species from northeastern Mexico. J. Appl. Anim. Res. 23: 33-41.

Mupangwa, J.F., T. Acamovic, J.H. Topps, N.T. Ngongoni, y H. Hamudikuwanda. 2000. Content of soluble and bound condensed tannins of three tropical herbaceous forage legumes. Anim. Feed Sci. Technol. 83:139-146.

Nagy, J.G. y Regelin, W.L. 1975. Comparision of digestive organ size of three deer species. Wildlife Journal of Management, 39: 621-624.

NAP. 2000. Dietary Reference Intakes for Thiamin, Riboflavin, Niacin, Vitamin B6, Folate, Vitamin B12, Pantothenic Acid, Biotin, and Choline.

Narjisee, H., M.A. Elhonsali y J.D. Olsen. 1995. Effect of oak (Quercus ilex) tannins on digestion and nitrogen balance in sheep and goats. Small Rumin. Res. 18:201-207.

National Research Council. 2001. Nutrient Requirements of Dairy Cattle, Seventh Revised Edition. National Research Council, National Academy Press,

Nelson, B.D., C.R. Montgomery, P.E. Schilling, and L. Mason. 1975. Effects of fermentation time on in vivo/in vitro relationships. J. Dairy Sci. 59:270-277.

Nelson, ME and Mech, LD. 1992. Dispersal in female white-tailed deer. Journal of Mammology. 4: 891-984.

Ngwa, A; Pone, DK; Mafeni, JM. 2000. Feed selection and dietary preferences of forage by small ruminants grazing natural pastures in the Sahelian zone of Cameroon. Animal Feed Science and Technology 88:253-266.

Nherera, F.V., Ndlovu, L. R., and Dzowela, B.H.R., 1999. Relationships between in vitro gas production characteristics, chemical composition and in vivo quality measures in goats fed tree fodder supplements. Small Rumin. Res. 31,117-126.

Nocek, J. E., and S. Tamminga. 1991. Site of digestion of starch in the gastrointestinal tract of dairy cows and its effect on mild yield and composition. J. Dairy Sci. 74:3598-3629.

Nordkvist, E., and P. Aman. 1986. Changes during growth in anatomical and chemical composition and in-vitro degradability of lucerne. J. Sci. Food Agric. 37:1-7.

Norton, B. W. and D. P. Poppi. 1994. Composition and Nutritional Attributes of Pasture Legumes, en: Tropical Legumes in Animal Nutrition. Editors J. P. F. D´Mello and C. Devendra. Cab International, Wallingford, Reino Unido. pp. 23 - 48.

NRC, 2007. National Research Council. Nutrient Requirement of Small Ruminant. Sheep, goats, cervids, and new world camelids. National Academy Press. Washinton, DC.

NRC. 2000. National Research Council. Nutrient Requirements of Domestic Animals: Nutrient Requirements of Beef Cattle, 8th edition. National Academic Press, Washington, DC., pp. 75-84.

NRC. 2000. National Research Council. Nutrient Requirements of Domestic Animals: Nutrient Requirements of Beef Cattle, 8th edition. National Academic Press, Washington, DC., pp. 90-98.

O´Reilly, G. 2002. Tannin wars. Department of Business, Industry & Resource Develop. pp. 234-255.

Ockenfels, R. A., D. E. Brooksy C. H. Lewis. 1991. General ecology of Coues white-tailed deer in the Santa Rita Mountains. Arizona Game & Fish Deparment Techican Report 6:1-73.

Odell, B.L. and R.A. Sunde. 1997. Introduction. In: Handbook of Nutritionally Essential Mineral Elements. B.L. Odell and R.A. Sunde (editors). Marcel Dekker Inc., Nueva York, pp. 1-12.

Ørskov, E.R. 1992. Protein Nutrition in Ruminants. 2nd edn. Academic Press, London, p. 175.

Ørskov, E.R. y Kay, R.N.B. 1987. Non-microbial digestion of forages by herbivores. En: J.B. Hacker y J.H. Ternouth (editors). The Nutrition of Herbivores. Academic Press, Inc, Orlando Fl. Estados Unidos. pp. 267-280.

Ørskov, E.R. y McDonald, I. 1979. The estimation of protein degradability in the rumen from incubation measurements weighed according to rate of passage. J. Agri. Sci. (Cambridge), 92: 499-503.

Owens, F.N. 1988. Ruminal Fermentation. En: The Ruminant Animal Digestive Physiology and Nutrition, Waveland Press, Inc. pp. 145-171.

Padh, H. 1991. Vitamin C: Newer insights into its biochemical functions. Nutrition Reviews. 49: 65-79.

Paolini, V. y H. Hoste. 2006. Effects of tannins in goats infected with gastrointestinal nematodes. BSAS Publication 34. The assessment of intake, digestibility and the roles of secondary compounds. Edited by C.A. Sandoval-Castro, F.D.DeB.D. Hovell, J.F.J. Torres-Acosta and A. Ayala-Burgos. Nottingham University Press. pp. 209-220.

Parissi, Z.M., T.G. Papachristou y A.S. Nastis. 2005. Effect of drying method on estimated nutritive value of browse species using an in vitro gas production technique. Anim. Feed Sci. Technol. 123-124:119-128.

Parker, K.L., M.P. Gillingham, T.A. Hanley and C.T. Robbins. 1999. Energy and protein balance of free-ranging black-tailed deer in a natural forest environment. Wildlife Monographs, No. 143. a Publication of The Wildlife Society, pp. 6-7.

Pekins, P. J., K. S. Smith y W. W. Mautz. 1998. The energy cost of gestation in white-tailed deer. Canadian Journal of Zoology 76:1091-1097.

Pell A N and Schofield P 1993 Computerized monitoring of gas production to measure forage digestion in vitro. Journal of Dairy Science. 76: 1063-1073.

Pereira-Filho, J.M., E.L. Vieira, A. Kamalak, A.M.A. Silva, M.F. Cezar y P.M.G. Beelen. 2005 Correlação entre o teor de tanino e a degradabilidade ruminal da matéria seca e proteína bruta do feno de jurema-preta (Mimosa tenuiflora Wild) tratada com hidróxido de sódio. Livestock Research for Rural Development.Volume 17, Art. #91.

Perevolotsky, A., S. Landau, N. Slanikove y F. Provenza. 2006. Upgrading tannin-rich forages by supplementing ruminants with Polyethilene Glycol (PEG). BSAS Publication 34. The assessment of intake, digestibility and the roles of secondary compounds. Edited by C.A. Sandoval-Castro, F.D.DeB.D. Hovell, J.F.J. Torres-Acosta and A. Ayala-Burgos. Nottingham University Press. pp. 221-234.

Perez-Correa, M.E., Vazquez de Aldana, B.R., Garrcia-Criado, B. and Garcia-Ciudad, A. Variations in nutricional quality and biomasa production of semiarid grasslands. J. Range Mange. 51:570-576.

Pichersky, E. y D.R. Gang. 2000. Genetics and biochemistry of secondary metabolites in plants: an evolutionary perspective. Trends in Plant Sci. 5:439-445.
Pineda, N; Pérez, E; Vásquez, F. 2009. Evaluación de la selectividad animal de plantas herbáceas y leñosas forrajeras durante dos épocas en la zona alta del Municipio de Muy Muy, Nicaragua. Agroforestería en las Américas No. 47:46-50.
Pins, R.A. 1991. the rumen ciliates and their functions. En: J.P. Jouany (editor). Rumen Microbial Metabolism and Ruminant Digestion. INRA, Francia, pp. 39-52.
Pollock, M. T., Whittaker, D.G., Demarais, S. and Zaiglin, R.F. 1994. Vegetation characteristics influencing site selection by male white-tailed deer in Texas. J. Range Manage., 47: 235-239.
Pond, W.G., D.C. Church and K.R. Pond. 1995. Basic Animal Nutrition and Feeding, fourth edition, John Wiley and Sons, Inc. Corvalis OR. Pp. 101-113.
Provenza, F.D. 2006. Behavioural mechanisms influencing use of plants with secondary metabolites by herbivores. BSAS Publication 34. The assessment of intake, digestibility and the roles of secondary compounds. Edited by C.A. Sandoval-Castro, F.D.DeB.D. Hovell, J.F.J. Torres-Acosta and A. Ayala-Burgos. Nottingham University Press. pp. 183-195.
Rabb, L., B. Cafantaris, T. Jilg, and K.M. Menke. 1983. Rumen protein degradation and biosynthesis. 1. A new method for determination of protein degradation in rumen fluid in vitro. British. Journal of Nutrition, 50:569.
Ralph, J., R.D. Hatfield, S. Quideau, R.F. Helm, J.H. Grabber, and H.G. Jung. 1994. Pathway of p-coumaric acid incorporation into maize lignin as revealed by NMR. J. Am. Chem. Soc. 116:9448.
Ramírez, R.G., G.F.W. Haenlein, A. Treviño and J. Reyna. 1996. Nutrient and mineral profile of white-tailed deer (Odocoileus virginianus, texanus) diets in northeastern Mexico. Small ruminant Research, 23: 7-16.
Ramírez Lozano R.G. 2009. Nutrición de Rumiantes: Sistemas Extensivos. Editorial Trillas, 2ª edición. pp. 52-74.
Ramírez Lozano, RG, GF Alanís Flores y MA Nuñez González. 2000. Importancia del nopal (Opuntia engelmannii) en la alimentación del ganado. Ciencia UANL. 3:267-273.
Ramírez, R.G. and M.A. Núñez-González. 2006. Chemical Composition, Digestion and Mineral Content of Native Forbs Consumed by Range Sheep. Journal of Animal and Veterinary advances. 5: 1158-1164
Ramírez, R.G. and M.A. Núñez-González. 2006. Chemical Composition, Digestion and Mineral Content of Native Forbs Consumed by Range Sheep. Journal of Animal and Veterinary advances. 5: 1158-1164.
Ramírez, R. G. 1989. Estudios nutricionales de las cabras en el noreste de México. Segunda parte. Cuaderno de Investigación No. 13. Dirección General de

estudios de Posgrado. Universidad Autónoma de Nuevo León. San Nicolás de los Garza, N. L., México. pp. 12-17.

Ramírez, R. G., Alanis-Flores, G. F., Núñez-González, M. A., 2000. Dinámica estacional de la digestión ruminal de la materia seca del nopal. CIENCIA-UANL. 3, 267-273.

Ramírez, R. G., G.F.W. Haenlein, A. Treviño y J. Reyna. 1996. Nutrient and mineral profile of whit-tailed deer (Odocoileus virginianus, texanus) diets in northeastern Mexico. Small Ruminants Research. 23:7-16.

Ramírez, R.G. 1999. Feed resources and feeding techniques of small ruminants under extensive management systems. Small Rumin Res. 34:215-230.

Ramirez, R.G. Qintanilla, J.B. and Aranda, J. 1997. Food habits of white-tailed deer (Odocoileus virginianus, texanus) in northeastern México. Small Ruminant Research 25:141-146.

Ramirez, R.G. y J.A. Lara. 1998. Influence of native shrubs Acacia rigidula, Cercidium macrum and Acacia farnesiana on digestibility and nitrogen utilization by sheep. Small Ruminant Research. 28:39-45.

Ramírez, R.G., G.F.W. Haenlein, C.G. García-Castillo, M.A. Núñez-González. 2003. Protein, lignin and mineral contents and in situ dry matter digestibility of native Mexican grasses consumed by range goats. Small Ruminant Research. (En Prensa).

Ramírez, R.G., J.B. Quintanilla and J. Aranda. 1997. White-tailed deer food habits in northeastern Mexico, Small ruminant Research, 25: 141-146.

Ramírez, R.G., Mireles, E., Huerta, J.M. and Aranda, J. 1995. Forage selection by range sheep on a buffelgrass (Cenchrus ciliaris) pasture. Small Ruminant Research, 17: 129-135.

Ramírez, R.G., Mireles, E., Huerta, J.M. and Aranda, J. 1995a. Forage selection by range sheep on a buffelgrass (Cenchrus ciliaris) pasture. Small Rumin Res. 17: 129-135.

Ramírez, R.G., R.R. Neira-Morales y J.A. Torres-Noriega. 2000a. Digestión ruminal de la proteína de siete arbustos nativos del nordeste de México. I. J. Experimental Botany (YTON), 67:29-35.

Ramírez, R.G., R.R. Neira-Morales, J.A. Torres-Noriega and AC Mercado-Santos. 2000b. Seasonal variation of chemical composition and crude protein digestibility in seven shrubs of NE Mexico. I. J. Experimental Botany (YTON), 67:77-82.

Ramírez, R.G., R.R. Neira-Morales, J.A. Torres-Noriega and AC Mercado-Santos. 2000b. Seasonal variation of chemical composition and crude protein digestibility in seven shrubs of NE Mexico. I. J. Experimental Botany (YTON), 67:77-82.

Ramírez, R.G., R.R. Neira-Morales, J.A. Torres-Noriega and AC Mercado-Santos. 2000b. Seasonal variation of chemical composition and crude protein digestibility in seven shrubs of NE Mexico. I. J. Experimental Botany (YTON), 67:77-82.

Ramírez, RG, L. Háuad, R. Foroughbackhch y JG. Moya-Rodríguez, 1998. Digestión ruminal de la proteína cruda del heno de alfalfa y de las hojas de 10 arbustos nativos del noreste de México. Ciencia UANL 1: 54-58.

Ramírez, RG, L. Háuad, R. Foroughbackhch y JG. Moya-Rodríguez, 1998. Digestión ruminal de la proteína cruda del heno de alfalfa y de las hojas de 10 arbustos nativos del noreste de México. Ciencia UANL 1: 54-58.

Ramírez, RG, L. Háuad, R. Foroughbackhch y JG. Moya-Rodríguez, 1998. Digestión ruminal de la proteína cruda del heno de alfalfa y de las hojas de 10 arbustos nativos del noreste de México. Ciencia UANL 1: 54-58.

Ramírez, RG, L. Háuad, R. Foroughbackhch y JG. Moya-Rodríguez, 1998. Digestión ruminal de la proteína cruda del heno de alfalfa y de las hojas de 10 arbustos nativos del noreste de México. Ciencia UANL 1: 54-58.

Ramírez-Lozano, R.G. 1989. Estudios Nutricionales de las Cabras en el Noreste de México: Primera parte. Dirección general de Estudios de Posgrado, UANL, San Nicolás de los Garza, N.L., México. Cuadernos de Investigación No. 6. p. 10.

Ramírez-Lozano, R.G. 1996. Manual de Técnicas de Investigación en Nutrición de Rumiantes en Pastoreo. Facultad de Medicina Veterinaria y Zootecnia, Universidad Autónoma de Nuevo León. pp. 18-20.

Ramírez-Lozano, R.G. 2004. Nutrición del Venado Cola Blanca. Publicaciones Universidad Autónoma de Nuevo León, San Nicolás de los Garza, N.L., México. pp. 173-188.

Ramírez-Lozano, R.G. 2006. Nutritional characteristics of browse species from Northeastern Mexico consumed by small ruminants. BSAS Publication 34. The assessment of intake, digestibility and the roles of secondary compounds. Edited by C.A. Sandoval-Castro, F.D.DeB.D. Hovell, J.F.J. Torres-Acosta and A. Ayala-Burgos. Nottingham University Press. pp. 251-260.

Ramírez-Lozano, R.G. 2007. Los Pastos en la Nutrición de Rumiantes. Publicaciones Universidad Autónoma de Nuevo León. pp. 25-37.

Ramírez-Lozano, R.G. 2009. Nutrición de Rumiantes: Sistemas Extensivos. 2ª Edición. Editorial Trillas, pp. 204-220.

Ramírez-Lozano, R.G. 2009. Nutrición de Rumiantes: Sistemas Extensivos. 2ª Edición. Editorial Trillas, pp. 204-220.

Rautenstrauch, K.R. & Krausman, P.R. (1989) Influence of water availability and rainfall on movements of desert mule deer. J. Mamm, 70, 197-201.

Reed, J.D. 1995. Nutritional toxicology of tannins and related polyphenols in forage legumes. J. Anim. Sci. 73:1516-1522.

Reed, J.D. 2001. Effects of proanthocyanidins on the digestion and analysis of fiber in forages. J. Range management. 54: 466-473.

Reid, R.I. Nitrogen components of forages and feedstuffs. 2000. En: J.M. Asplund (Editor). Principles of Protein Nutrition of Ruminants. CRC Press, Boca Raton, FL, EUA, pp. 43-70.

Richardson CL. 1999. Factors affecting deer diets and nutrition. South Texas Rangelands. Texas Agricultural Extension Service, Texas A and M University. L-2393 pp. 2-3.

Richarson, C.L. 1999. Factors affecting deer diets and nutrition. South Texas Rangelands. Texas Agricultural Extension Service. Texas A & M University, Colledge Station. I. 2393, pp. 1-6.

Rittner, U. y J.D. Reed. 1992. Phenolics and in vitro degradability of protein and fibre in West African browse. Journal of the Science of Food and Agriculture, 58: 21-28.

Robbins, C.T. 2001. Wildlife Feeding and Nutrition. 2a Edición. Academic Press, New York. EUA. pp.

Rodríguez C.A., E. Valencia 2008 Microbiología Ruminal. Ruminantia 3:1 disponible en http://www.uprm.edu/agricultura/inpe/ruminantia/ruminantia3-1-2008.pdf accesado 02/09/2009.

Rodríguez Lozano J. E. 1995. Dinámica estacional del contenido de minerales en el forraje de pastos nativos y hierbas nativas edibles del Estado de Nuevo León. Tesis Profesional, Facultad de Medicina Veterinaria y Zootecnia, Universidad Autónoma de Nuevo León, México.

Rogosic, J., J.A. Pfister, F.D. Provenza y D. Grbesa. 2006. Sheep and goat preference for and nutritional value of mediterraneasn maquis shrubs. Small Rumin. Res. 64:169 -179.

Rogosic, J., R.E. Estell, S. Ivankovic, J. Kezic y J. Razov. 2008. Potential mechanisms to increase shrub intake and performance of small ruminants in Mediterranean shrubby ecosystems. Small Rumin. Res. 74:1 -15.

Rogosic, J., S.R. Moe, D. Skobic, Z. Knezovic, I. Rozic, M. Zivkovic y J. Pavlicevic. 2009. Effect of supplementation with barley and activated charcoal on intake of biochemically diverse Mediterranean shrubs. Small Rumin. Res. 81:79-84.

Rohweder, D.A., R.F. Barnes, and N. Jorgensen. 1978. Proposed hay grading standards based on laboratory analyses for evaluating quality. J. Anim. Sci. 47:747-759.

Romero, L.C.E., G.J.M. Palma y J. López. 2000. Influencia del pastoreo en la concentración de fenoles y taninos condensados en Gliricidia sepium en el trópico seco. Livestock Research for Rural Development 4:1-9 http://www.cipav.org.co/lrrd/lrrd12/4/rome124.htm.

Romero-Paredes Rubio, J.I. y Ramírez Lozano, R.G. 2003. Atriplex canescens (Purch, Nutt) como fuente de alimento para las zonas áridas. Ciencia UANL, 5: 85-92.

Rosenstock, S.S., Ballard, W.B. & deVos, J.C. second (1999) Benefits and impacts of wildlife water developments. J. Range Manage, 52, 302-311.
Roubroeks, J.P., Andersson, R., Mastromauro, D.I., Christensen, B.E. and Åman, P. 2001. Molecular weight, structure and shape of oat (1?3),(1?4)-b-D-glucan fractions obtained by enzymatic degradation with (1?4)-b-D-glucan 4-glucanohydrolase from Trichoderma reesei, Carbohydrate Polymers. 46: 275-285.
Rowell, A., Deyer, J., Hofmann, R.R., Lechner-Doll, M., Meyer, H.H.D., Shirazi-Beechey, S.P. and Streich, W.J. 1999. Abundance of intestinal Na+glucose cotransporter (SGLT1) in roe deer, J. of animal Physiology Nutrition, 82: 25-32.
Ruckstuhl, K. E., P. Neuhaus. 2002. Sexual segregation in ungulates: a comparative test of three hypotheses. Biology Review. 77, Pp.77-96.
Russell, F.L., Zippin, D.B. y Fowler N.J.. 2001. Effects of White-tailed Deer (Odocoileus virginianus) on Plants, Plant Populations and Communities: A Review. The American Midland Naturalist 146:1-26.
Russell, R.W. and S.A. Gahr. 2000. Glucose availability and associated metabolism . In: Farm Animal Metabolis and Nutrition. Edited by J.P.F. D'Mello. CABI, Publishing, Reino Unido.
Rzedowski, J. 1986. Vegetación de México. Editorial Limusa. Escuela Nacional de Ciencias Biológicas. Instituto Politécnico Nacional.
Sánchez-Rojas, G. 2004. ¿Qué es la segregación sexual y como se ha explicado? Memorias. IX Simposio de Venados en México FMVZ-UNAM. Pachuca, Hidalgo.
Santos Domínguez, J.L. 1995. Perfil nutritivo y digestibilidad in situ de la proteína cruda de ocho hierbas nativas de Nuevo León, Colectados en Primavera. Tesis Profesional, Facultad de Medicina Veterinaria y Zootecnia, Universidad Autónoma de Nuevo León.
Schaefer, J y Marin, MB. 2001. White-tailed deer of Florida. WEC-133, Florida Cooperative Extension Service, Institute of food and Agricultural science, University of Florida. pp 1-11.
Schofield, P. 2000. Gas Production Methods. In: Farm Animal Metabolism and Nutrition. Wallingford (UK). CAB International. 450 p.
Scott, J.M. 1999. Folate and vitamin B12. Proceedings Nutrition Society. 2 : 441-8.
Scout, M.L., Nesheim, M.C. y Young, R.J. 1982. Nutrition of the chiken. Tercera edición, M.L. Scott y Associados, Ithaca, NY, p. 562.
Servello, F.A. y Webb, K.E. 1987. Predicting the metabolizable energy in the diet of ruffed grouse, J. Wildl. Manage. 51: 560-567.
Sfeinvar, L. 1982. La Familia de las Cactaceas en el Valle de México, Tesis, Doctor en Biología, UNAM, México, pp 31, 485 y 493.

Shin, E.T., Hudson, R.J., Gao, X.H. y Suttie, J.M. 1998. Nutritional requirements and management strategies for farmed deer. En: Prodeedings of 8th World Conference on Animal Nutrition, Seul, Corea pp. 459-475.

Singh, B., A. Sahoo, R. Sharma y T.K. Bhat. 2005. Effect of polyethylene glycol on gas production parameters and nitrogen disappearance of some tree forages. Anim. Feed Sci. Technol. 119:23-124.

Smith, P. 1991. Odocoileus virginianus. Mammalian Species, 388: 1-13.

Sollenberger, L.E., and J.C. Burns. 2001. Canopy characteristics, ingestive behaviour and herbage intake in cultivated tropical grasslands. p. 321-327. In J.A. Gomide et al. (ed.) Proc. Int. Grassl. Cong., 19th, São Pedro, SP, Brazil. 10-21 Feb. 2001, Brazilian Soc. Anim. Husb., Piracicaba, SP, Brazil.

Sosa, R.E.E., R.D. Pérez, R.L. Ortega y B.G, Zapata. 2004. Evaluación del potencial forrajero de árboles y arbustos tropicales para la alimentación de ovinos. Técnica Pecuaria en México. 42:129-144.

Spaeth, D. F., K. J. Hundertmark, R. T. Bowyer, P. S. Barboza, T. R. Stephenson y R.O. Peterson. 2001. Incitor arcades of Alaskan moose: is dimorphism related to sexual segregation? Alces 37:217-226.

Sparks, D.R. y Malechek, J.C. 1968. Estimating percentage of dry weight in diets using a microscope technique. J. Range Management 21: 264-270.

Spears, J.W. 1998. Reevaluation of metabolic essentiality of minerals. Proceedings: New Technologies for the Production of "Next Generation" Feeds and additives. The 8th World Conference on animal Production, Seoul National University, Seoul Korea, pp. 68-77.

Staaland, H., R.G. White, J.R. Luick and D.F. Holleman. 1980. Dietary influences of sodium and potassium metabolism of reindeer. Canadian J. Zoology, 58: 1728-1734.

Stewart, C.S. 1994. Plant-animal and microbial interactions in ruminant fibre degradation. p. 13-28. En: Microorganisms in ruminant nutrition. R.A. Prins y C.S. Stewart (eds.), Notting-ham University Press, Nottingham.

Stewart, C.S. 1994. Plant-animal and microbial interactions in ruminant fibre degradation. p. 13-28. En: Microorganisms in ruminant nutrition. R.A. Prins y C.S. Stewart (eds.), Notting-ham University Press, Nottingham.

Stewart, K.M., Fulbrith, T.E. y Drawe, D.L. 2000. White tailed deer use of clearings relative to forage availability. Journal of Wildlife Management. 64:733-741.

Strickland, B. K., D. G. Hewitt, C. A. Deyoung YR. L. Bingham. 2005. Digestible energy requirements for maintenance of body mass of white-tailed deer in southern Texas. Journal of Mammalogy 86:56-60

Sultan, S., Negi, A.S., Agarwal, D.K., Katiyar, P.K. y Singh, U.P. 2000. Chemical composition, in vitro dry matter digestibility, total phenolics and proanthocyanidins in hedge lucerne (Desmanthus virgatus) exotic germplasm. Indian Journal of Animal Sciences, 70: 1246-1249.

Swain, R. A., J. V. Nolan, y A. V. Klieve.1996. Natural variability and diurnal fluctuations within the bacteriophage population of the rumen. Appl. Environ. Microbiol. 62:994-997.
Terashima, N., K. Fukushima, L.-F. He, and K. Takabe. 1993. Comprehensive model of the lignified plant cell wall. In: H.G. Jung, D.R. Buxton, R.D. Hatfield, and J. Ralph (Ed.) Forage Cell Wall Structure and Digestibility. p 247. ASA-CSSA-SSSA, Madison, WI.
Theander, O., and P. Aman. 1980. Chemical composition of some forages and various residues from feeding value determinations. J. Sci. Food Agric. 31:31.
Tilley, J.M.A. y Ferry, R.A. 1963. A two satage technique for the in vitro digestion of forage crops. J. Br. Grassl. Soc. 18: 104-108.
Tomlinson, D.J. 2002. Effect of organic and inorganic trace mineral suppmentantion on beef and dairy production. Prentacion de la XXX Reunión Anual de la Asociación Mexicana de Producción Animal, Octubre 13-15, Guadalajara Jalisco.
Torres-Acosta, J.F., M.A. Alonso-Díaz, H. Hoste, C.A. Sandoval-Castro. y A.J. Aguilar-Caballero. 2008. Efectos negativos y positivos del consumo de forrajes ricos en taninos en la producción de caprinos. Tropical and Subtropical agroecosystems. 8:83-90.
Underwood, E.J. and N.F. Shuttle. 1999. The Mineral Nutrition of Livestock, 3rd edn. CAB International, Wallingford. Reino Unido.
Ushida, K., & Jouany, J.P. 1985. Effect of protozoa on rumen protein degradation in sheep. Reproduction, Nutrition, Dévelopment, 25: 1075-1081.
Van Dijk CJ, DC Lourens. 2001. Effects of anionic salts in a prepartum dairy ration on calcium metabolism. J S Afr Vet Assoc 72, 76-80.
Van Soest, P. J., J. B. Robertson, B. A. Lewis. 1991. Methods for dietary fiber, neutral detergent fiber, and nonstarch polysaccharides in relation to animal nutrition. J. Dairy Sci. 74:3583-3597.
Van Soest, P.J. 1993. Cell wall matrix interactions and degradation session synopsis. In: H.G. Jung, D.R. Buxton, R.D. Hatfield, and J. Ralph (Ed.) Forage Cell Wall Structure and Digestibility. p 377. ASA-CSSA-SSSA, Madison, WI.
Van Soest, P.J. 1994. Nutrional Ecology of the Ruminant. (2nd Edition). Comstock, Cornell University Press, Ithaca, NY.
Van Straalen, W.M. and Tamminga, S. 1990. Protein degradation of ruminant diets. In: Wiseman, J. and Cole, D.J.A. (eds) Feedstuff Evaluation. Butterworths, London, pp. 55-77.
Vargas-López V. R. 1991. Estudio morfoanatomico de las especies leñosas de la familia Leguminosae del estado de Nuevo León y su relación con la taxonómia. Tesis de Maestría. Facultad de Ciencias Biológicas, Universidad

Autónoma de Nuevo León, San Nicolás de los Garza, Nuevo León, México, pp. 53-54.
Varner, J.E., and L.S. Lin. 1989. Plant cell wall architecture. Cell. 56:233-239.
Varner, L.W. and Blankenship, L.H. 1987. South Texas Shrubs: nutritive value and utilisation by herbivores. USDA, For. Ser. Gen. Tech. Rep. INI-222, p.7.
Varner, LW. and HG Hughes. 1999. Nutrition Effects on Fawn, Doe and Buck Deer. Wildlife Management Handbook II-B. Texas Agricultural Experiment Station, Wildlife and Fisheries Sciences Department, Texas A&M University System. pp. 7-10.
Vázquez-Rodríguez, M., Ruiz de León, M.T., Valdés-Reyna, J. y López-Trujillo, R. 1985. Características morfológicas de especies forrajeras del matorral desértico Micrófilo en el noreste de México. Folleto de Divulgación Vol. 1, No. 6, Universidad Autónoma Agraria Antonio Narro. pp. 2-13.
Velásquez, R; Pezo, D; Skarpe, C; Ibrahim, M; Mora-Delgado, J; Benjamin, T. 2009. Selectividad animal de forrajes herbáceos y leñosos en pasturas seminaturales en Muy Muy, Nicaragua. Agroforestería en las Américas No. 47:51-60.
Ventura, M.R., J.I.R. Castañón, M.C. Pieltain y M.P. Flores. 2004. Nutritive value of forage shrubs: Bituminaria bituminosa, Rumex lunaria, Acacia salicina, Cassia sturtii and Adenocorpus foliosus. Small Rumin. Res. 52:13-18.
Villareal Espino, O. A. y Marín, M.M. 2005. Fuentes agua de origen vegetal para el venado cola blanca mexicano. Archivos de zootecnia. 54: 191-196
Waghorn, G. 2008. Beneficial and detrimental effects of dietary condensed tannins for sustainable she-ep and goat production-Progress and challenges. Anim. Feed Sci. and Techno., 147:116-122.
Waghorn, G.C. y W.C. McNabb. 2003. Consequences of plant phenolic compounds for productivity and health of ruminants. Proc. Nut. Soc. 62:383-392.
Walker, N.D., Newbold, C.J. y Wallace, R.J. 2005. Nitrogen metabolism in the rumen. En: E. Pfeffer y A.N. Hristov (editors). Nitrogen and Phosphorous Nutrition of Cattle. CABI Publishing. pp. 71-100.
Wallace, R.J. Amino acid and protein synthesis, turnover, and breakdown by ruminal microorganisms. 2000. En: J.M. Asplund (Editor). Principles of Protein Nutrition of Ruminants. CRC Press, Boca Raton, FL, EUA, pp. 71-112.
Wallmo, O. 1981. Mule and black-tailed deer of North America. A Wildlife Management Institute Book.
Warnock, B. H. 1970. Wildflowers of the Big Bend Country, Texas. Sul Ross State Univ., Alpine, Texas. EUA., p.157.
Washington, DC.
Waterman, P.G., 2000. The tannins-An overview. In: Tannins in livestock and human nutrition. Proceedings of an International Workshop. Editor J.D.

Brooker. Australian Centre for International Agricultural Research, Canberra, Australia. pp. 10-13.

Waterman, R.C., Grings, E.E., Geary, T.W., Roberts, A.J. Alexander, L.J. and MacNeil, M D. 2007. Influence of seasonal forage quality on glucose kinetics of young beef cows. Journal of Animal Science. 85:2582-2595.

Watkins, B. E.; J. H. Witham, D. E. Ullrey, D. J. Watkins y J. M. Jones. 1991. Body composition and condition evaluation of white-tailed deer fawns. J. Wildl. Manage. 55:39-51.

Webb, K.E. and Bergman, E.N. 1991. Amino acid and peptide absorption and transport across the intestine. In: Tsuda, T., Sasaki, Y. and Kawashima, R. (eds) Physiological Aspects of Digestion and Metabolism in Ruminants. Academic Press, San Diego, pp. 111-128.

Webster, A.J.F. 1996. The metabolizable protein synthesis for ruminants. In: Gransworthy, P.C. and Cole, D.J.A. (eds) Recent Developments in Ruminant Nutrition- 3. Nottingham University Press, Nottingham, pp. 55-70.

Weeks, H.P., Jr. y C.M. Kirkpatrick. 1976. Adaptations of whitetailed deer to naturally occurring sodium deficiencies. J. Wildlife Management. 40: 474-477.

Werkely, W. F. 1994. Selective feeding by black-tailed deer: Forage quality or abundance? Journal of Mammalogy 75:905-913.

Weston, R.H. 1982. Animal factors affecting feed intake. p. 183-198. In J.B. Hacker (ed.) Nutritional limits to animal production from pastures. CAB Intl., Slough, UK.

White, B.A., R.I. Mackie, and K. C. Doerner. 1993. Enzymatic hydrolysis of forage cell walls. p. 455-484. In H.G. Jung, D.R. Buxton, R.D. Hatfield, and J. Ralph (ed.) Forage Cell Wall Structure and Digestibility. ASA-CSSA-SSSA, modison, WI.

Whitehead, D.C. 2000. Nutrient Elements in Grassland, soil-plant-animal-relationships. CABI Publishing, pp. 70-93.

Williams, B. A. 2000. Cumulative gas-production techniques for forage evaluation. In: Givens D I, Owen E, Omed H M and Axford R F E (editors). Forage Evaluation in Ruminant Nutrition. Wallingford (UK). CAB International. 475 p.

Wisdom, C.S., Gonzalez-Coloma, A. y Rundel, P.W. 1987. Ecological tannins assays. Evaluation of Proanthocyanidins, protein binding assays and protein precipitation potential. Oecology, 72: 395-401.

Woodward, A., y J. D. Reed. 1997. Nitrogen Metabolism of sheep and goats consuming Acacia brevispica and Sesbania sesban. J. Anim. Sci. 75:1130-1137.

Wu G. and Self, J.T. 2005. Proteins. En: W.G. Pond y A.W. Bell (Editores). Encyclopedia of Animal Science. Marcel Dekker, NY, EUA. pp. 323-325.

Wyburn, R.S. 1980. Digestive Physiology and Metabolism in ruminants. Editores Y. Ruckebusch y P. Thivend. MTP Press Ltd., Lancanter, UK.

Yagil, R., Amir, H., Abu-Rabiya, Y. & Etzion, Z. (1986) Dilution of Milk: a physiological adaptation of mammals to water stress?. J. Arid Env, 11, 243-249

Yerex, D. and I. Spiers. 1987. Modern Deer Farm Management. Ampersand Publishing Ltd. Carterton. New Zealand. Pp. 49-68.

Yokoyama, M.T. y Johnson, K.A. 1988. Microbiology of the rumen and intestine In: Church, DC. (ed.). The ruminant Animal: Digestive Physiology and Nutrition. Prentice Hall. Engelwood Cliffs, New Jersey, pp. 125-144.

www.ingramcontent.com/pod-product-compliance
Lightning Source LLC
Chambersburg PA
CBHW021405210526
45463CB00001B/229